Worker Protection, Japanese Style

Cornell International Industrial and Labor Relations
Report Number 21

Worker Protection, Japanese Style

Occupational Safety and Health in the Auto Industry

Richard E. Wokutch

ILR Press
Ithaca, New York

Cover design by Marcelle Toor.

Library of Congress Cataloging-in-Publication Data

Wokutch, Richard E.
Worker protection, Japanese style : occupational safety and health in the auto industry / Richard E. Wokutch.
p. cm.—(Cornell international industrial and labor relations report ; no. 21)
Includes bibliographical references and index.
ISBN 0-87546-186-7 (alk. paper)—ISBN 0-87546-187-5 (alk. paper : pbk.)
1. Automobile industry and trade—Japan—Safety measures.
2. Automobile industry workers—Health and hygiene—Japan.
3. Automobile industry and trade—United States—Safety measures.
4. Automobile industry workers—Health and hygiene—United States.
I. Title. II. Series: Cornell international industrial and labor relations reports ; no. 21.
HD7269.A82J38 1992
363.11'96292'0952—dc20 91-48201

Copies may be ordered through bookstores
or directly from
ILR Press
School of Industrial and Labor Relations
Cornell University
Ithaca, NY 14853-3901

Printed on acid-free paper in
the United States of America
5 4 3 2 1

With love and gratitude to my parents,
Charles and Annette Wokutch

Contents

Tables and Figures ix

Acknowledgments xi

1. Introduction 3

2. The Environmental Context for Occupational Safety and Health Activities in Japan 25

3. The Safety and Health System in Japan 55

4. Occupational Safety and Health at Jidosha 95

5. Occupational Safety and Health at Jidosha USA 145

6. Work Injury and Illness Experiences in the United States and Japan 196

7. Conclusions and Recommendations 224

Notes 237

References 245

Index 255

Tables and Figures

Tables

2-1. Total Workdays Lost Because of Industrial Disputes, 1975–88, in the United States and Japan, 43
3-1. Staffing of OSHA, 83
3-2. Labor Standards Bureau Inspection Activities and Staffing, 85
3-3. OSHA and State Program Inspection Activities and Staffing, 87
3-4. OSHA and State Program Enforcement Activities, 89
3-5. Percentage of OSHA Initial Inspections Resulting in Citations and Penalties, 90
3-6. Labor Standards Bureau Enforcement Activities, 91
4-1. Jidosha Injury and Illness Experience, 122
4-2. Injury and Illness Experiences of the Japanese Auto Industry, 124
4-3. Injury and Illnesses at Jidosha Subcontractors, 1985, 128–31
4-4. Number of Injury and Illness Cases and Frequency Rate by Categories of Subcontractors and Year, 132
5-1. Injury and Illness Experiences of Jidosha USA and Jidosha Japan, 190
6-1. Work Injury and Illness Experiences in the United States and Japan, 206
6-2. Adjusted Work Injury and Illness Experiences of the Private Sector in the United States and Japan, 210
6-3. Adjusted Work Injury and Illness Experiences of the Manufacturing Sector in the United States and Japan, 211
6-4. Adjusted Work Injury and Illness Experiences of the Automotive Industry in the United States and Japan, 212
6-5. Occupational Injury and Illness Rates by Establishment Size in the United States and Japan, 215

6-6. Injury and Illness Experiences of U.S. Subsidiaries of Japanese Automotive Firms, 218
6-7. U.S. Motor Vehicle Industry Injury and Illness Experiences of U.S. Plants Employing More Than 2,500 Employees, 222

Figures

1-1. Corporation-Stakeholder Relations, 14
1-2. Conflict-Coping Behavior, 19
3-1. Organization of the Labor Standards Bureau, 60
3-2. Character of Conflict-Coping Behavior by Management toward the Government, 80
3-3. Character of Conflict-Coping Behavior by Management toward Labor, 81
4-1. Jidosha and Industry Leave Accident Frequency Rate, 123
4-2. Jidosha and Industry Severity Rate, 125
6-1. Frequency Rate in Japan and the United States (All Industries), 207
6-2. Lost-Work Frequency Experiences of Japanese Automotive Firms, 220
6-3. Lost-Work Severity Experiences of Japanese Automotive Firms, 221

Acknowledgments

This book is an extension of work begun in the early 1980s when I received a Fulbright research fellowship to West Germany. My intent when I started that research was to study the regulation and management of corporate social performance in several countries whose orientation to labor-management-government relations was different from that in the United States. Occupational safety and health was an obvious choice as the issue on which to focus my research because cooperation among these parties seemed logical and highly beneficial, yet conflict among them is common in the United States. The automotive industry, which provides the industrial context for this study, is one that has always intrigued me and one whose importance in the world economy is enormous.

While I was in Europe I visited both U.S. and European automotive plants in several countries, including West Germany, Sweden, and the United Kingdom. I was then able to compare safety and health practices in these plants with safety and health practices in U.S. and European plants in the United States in light of the regulatory environments in the various countries. A Third World perspective was provided by a component of this research conducted in Kenya.

Although I had not initially planned to write a book about this research, it soon became clear that a book would be the best vehicle for communicating these results. It was at this point that my colleagues at Virginia Tech pointed out that occupational safety and health in Japan, which by then had become the number one auto-producing country in the world, was an important omission in my research. Shortly thereafter, I applied for and was very fortunate to receive a second Fulbright research fellowship, this one to Japan.

The Fulbright Program of the Japan–United States Educational Commission opened a great many doors in industry, government, unions, private agencies, and academic institutions in Japan. In fact, within a few weeks I realized that in an attempt to fill a gap in my research, I was accumulating far more information than I could include in one book. I concluded that book (Wokutch 1990a) with an overview of safety and health in Japan and decided to develop the analysis further. This book evolved from that research. Fulbright Program administrators and staff whose assistance on this project I would especially like to acknowledge include Caroline A. Mantano Yang, Masatoshi Ikeda, Kazuko Kamimura, Atsuko Ozawa, Mizuho Iwata, and Etsujiro Sakuma.

I am also indebted to my host institution in Japan, the Hiroshima Institute of Technology. The administration, faculty, staff, and students made great efforts to make me feel welcome and to assist me in my work. Because of my very limited skills in Japanese, many normally routine aspects of research and daily life became major obstacles that I could not have overcome without a great deal of help. Special thanks in this regard go to Tsuyoshi Ishida, Toshiro Kobayashi, and Bob McWilliams.

For reasons of confidentiality, I cannot thank by name individuals in the Japanese auto company on which this book focuses. Nevertheless, I would like to express my appreciation to them for the cooperation and hospitality they extended to me in Japan and the United States. A considerable organizational commitment was required to meet my endless requests for information. Although I did not spend nearly as much time at any of the other automotive firms I visited, I would like to thank the officials at these various Japanese and American firms.

Numerous other organizations and individuals in Japan helped me in my research. I wish to acknowledge the assistance and cooperation provided by the Japanese Ministry of Labor, especially Tadashi Nakamura, Seiji Machida, Hirohito Iino, Katutosi Tukamoto, and Tasuku Tadano; the Japan Industrial Safety and Health Association, especially Osamu Tomihisa and Yoko Yoshida; the Confederation of Japanese Automobile Workers' Unions, especially Masayuki Naoshima; the Research Institute of Industrial Safety; the Japanese Automobile Manufacturers' Association; the Japan Institute of Labour, especially Takeaki Urao, Osamu Hirota, and Masako Unoura; the U.S. Embassy in Tokyo, especially John Warnock; the Commonwealth of Virginia, Department of Economic Development, Tokyo office, especially Kazuaki Kobayashi; the University of Occupational and Environmental Health, especially Masaharu Kumashiro; the Institute for the Science of Labour, especially Yukio Endo, Kenji Nakaaki, and Takeshi Fujimoto; the National Institute of Industrial Health, especially Katsunori Homma; the Japan External Trade Office, especially Tsugio Kusajima; the National Defense Council for Victims of Karōshi, especially Hiroshi Kawahito; and the Japanese trade union Sōhyō, especially Chuji Shinta and Tsutomi Arai.

Scholars and other professionals in Japan, including Takahara (Koji) Matsuda and Michio Nitta at Musashi University, Yanosuke Mafune at Nihon University, Yasuhiko Matsuda at Yokohama National University, and Hiroshi Iwata, Tsuguo Kaneko, and Satoshi Kamata, provided important information and many helpful comments.

Without the assistance of interpreters, I could not have gotten very far in my research. Two of the many fine interpreters who helped me were Hitomi Nagayama and Michiko Minagawa.

Several organizations and individuals in the United States also provided considerable assistance with this project. These include the Washington, D.C. Embassy of Japan, especially Hideyo Nakano, Takanobu Teramoto, and Akira Oishi; the U.S. Bureau of Labor Statistics, especially Ethel Jackson, Betsy Jordan, James Barnhardt, and Arthur Neef; the Library of Congress, especially Takeo Nishioka and Yoshiko Yoshimura; and the United Auto Workers, especially Franklin Mirer.

I wish to acknowledge the support of the R. B. Pamplin College of Business at Virginia Tech which enabled me to take a research leave and to participate in summer research grant programs. Additional support was provided by the Barringer Center at Virginia Tech. Colleagues at Virginia Tech also provided a great deal of assistance. These include Robert Litschert, my department head when I started this project, and Jon Shepard, my department head when I finished, William Lehrman, Larry Alexander, Dow Scott, Kent Murrmann, Dane Partridge, Terry Cobb, Joe Cheng, and Deborah Milly.

My graduate assistants during this project provided me with invaluable assistance: Josetta McLaughlin, Liz McKinney, Bong-Gyu Park, Peggy Cloninger, and Craig Short. Other students who assisted with this project include Tamaki Kurusu, Mariko Reynolds, Hiromi Tachida, Yoko Kobayashi, Doug Spadaro, and Lynn Grimsley.

Secretarial support was also extensive. Alice LoMascolo was my secretary during much of this project. Other secretaries who worked with me were Sandy Crigger, Dinah Akers, Melissa Kessinger, Anna LoMascolo, Angie Beane, Frank Smith, and Sarah Wall.

I would also like to express my appreciation to the following individuals at other academic institutions who provided me with suggestions on this project: David Vogel, T. J. Pempel, Alfie Marcus, David McCaffrey, Edgar McVoy, Robert Masuda, Koiji Taira, Robert Asher, and Sanford Thatcher.

I would also like to thank the staff at the ILR Press, especially Erica Fox, and their reviewers for their assistance.

Most of all, I would like to thank my wife, Mary Ellen, and my son,

Andreas, for their support. The prolonged separations to conduct the field research and the late nights and work-filled weekends needed to complete the project required a great deal of patience.

Finally, I must note that although I received information, suggestions, and comments from numerous sources, the opinions expressed in this book and any errors or omissions are my own.

Worker Protection, Japanese Style

1

Introduction

Japan's economy and management practices have become objects of fascination in the West. Hundreds of books and articles have been published on the subjects, and consultants throughout the United States instruct business people in how to manage Japanese style. Unlike many other management practices and trendy economic theories, however, the interest in Japan does not seem to be a passing fancy.

Japan's economic progress in the wake of the devastation of World War II is indeed extraordinary. The economy has grown from a level so low that the Japanese needed American aid to prevent massive starvation to what is now the second largest economy in the world. By the end of the 1980s, Japanese workers, on average, earned almost as much as their American counterparts, and in industries such as automobiles, consumer electronics, optical equipment, robotics, and banking, Japan had become the undisputed world leader.

These economic and industrial successes have given the Japanese enormous international clout, which only recently has been recognized in the United States. In the mid–1980s Japan replaced the United States as the world's largest lender, and in just a few short years the United States became the nation with the largest debt. Japanese companies with abundant cash reserves made massive direct investments of capital in the United States, generating jobs, stimulating the economy, and providing models of efficiently run operations. These investments also raised concerns, however, about Japan's influence on the U.S. economy (Holstein 1986; Lehner and Murray 1990; Prestowitz 1989; Smith 1990). Already more than 400,000 Americans work for Japanese companies in the United States (Sherman 1991), and it is not unreasonable to expect, as some business forecasters have warned, that "your next boss may be Japanese." As I hope to make clear, this would be a mixed blessing.

In attempting to explain Japan's economic success, scholars have focused on such macroeconomic factors as its industrial policies, managed international trade, and cooperative government-industry relations, as well as more micro-oriented management practices. Among the latter are the management of productivity and quality, the just-in-time inventory system, participative management, lifetime employment, and, more generally, cooperative labor-management relations. Largely unexplored are the regulation and management of occupational safety and health and, more generally, issues pertaining to corporate social responsibility.

The purpose of this book is to examine these neglected issues, including Japan's institutional arrangements and practices for protecting workers on the job as well as its performance vis-à-vis occupational safety and health regulatory and managerial systems. A related purpose is to compare the national style of worker protection in Japan with that in the United States.[1] These issues are important in and of themselves and also for the insights they provide into more general managerial and regulatory practices and relations among labor, management, and government in Japan and in the United States.

Two conflicting myths about occupational safety and health in Japan have developed over the years. Neither is completely true, but, like most myths, there is a kernel of truth in each of them. The first myth is that Japanese corporations and the Japanese economy as a whole have built their considerable success on the backs of abused workers. This is a popular view among certain Western corporate and union leaders in search of excuses for their firms' and workers' relatively poor performance in recent years vis-à-vis their Japanese competition. This view has considerable historical validity, originating in the dismal working conditions in Japan during the early postwar years (Halberstam 1986; McVoy 1985).

The notion that current working conditions are excessively harsh was given credence in an exposé of the plight of the Japanese worker that received considerable media attention in the United States (Kamata 1982). Activities of the National Defense Council for Victims of Karōshi (Japanese for death from overwork), including publication of the book *Karōshi: When the "Corporate Warrior" Dies* (1990), have focused international attention on the problem of excessive work in Japan. Left-of-center political parties and labor unions in Japan have also directed attention to the problem.

The second myth is that the Japanese firm is a worker's paradise. Although they typically do not explicitly address occupational safety and health issues, books and articles that praise Japanese management invariably attribute Japan's success, at least in part, to its favorable treatment of workers and its cooperative labor-management relations (see, for example, Abegglen and Stalk 1985; Imai 1981; McMillan 1985; Monden 1983; Ouchi 1981; Pascale and Athos 1981; Schonberger 1982; Shimada 1980.) It is my hope that this book will get beyond the myths while examining the kernels of truth in each of them.

I will consider the regulation and management of occupational safety and health in Japan with particular reference to the automotive industry and, more particularly, one firm that recently established operations in the United States. Comparisons of the company's occupational safety and health activities in Japan and the United States and the results of these activities at national, industry, corporate, and plant levels reveal a great deal about the management of workplace safety and health in both countries and how the best national and corporate practices of each could be adapted cross-nationally. Such adaptations could potentially benefit both U.S. and Japanese industry and workers. I will also examine the degree to which each country's safety and health practices are culturally specific and whether national constraints place limits on the extent to which these practices are transferable.

Corporate social responsibility in Japan, another focus of this study, is fascinating for a number of reasons. Notable among these is that Japanese firms have been undeniably successful in managing many of their social responsibilities (whether they are referred to in this way or not). Also intriguing are the Japanese traditions of honor, shame, harmony, and responsibility to one's group(s), which influence corporate social responsibility from the level of the chief executive officer to that of the production worker. The clear-cut predominance of the interests of employees and the nation over the interests of other corporate stakeholders (e.g., stockholders, consumers, creditors, suppliers, the surrounding community, and so on) is yet another provocative aspect of corporate social responsibility in Japan, especially as it relates to occupational safety and health.

Importance of Occupational Safety and Health

Occupational safety and health is an important issue for both economic and social reasons. Public attention has been drawn to workplace safety and health in recent years because of such tragedies as the Union Carbide disaster in Bhopal, India; the explosions at the Arco and Phillips Petroleum chemical plants in Texas and at the Imperial Food Products chicken-processing plant in North Carolina; the Occidental Petroleum oil platform fire in the North Sea, and the *Exxon Valdez* oil spill; but a great many other less publicized incidents and conditions also resulted in work injuries that cumulatively had enormous economic and social consequences. The National Safety Council (1990) estimated that there were 1.8 million disabling work-related injuries and 10,600 work-related deaths in 1988, resulting in more than $47 billion direct and indirect costs. In recent years in the United States, the rate of workdays lost because of occupational injuries has typically exceeded that of workdays lost because of industrial disputes. Occupational illnesses result in even greater losses, although statistics are less precise because of the long gestation periods and uncertain causes of many illnesses. The Office of Technology Assessment of the U.S. Congress (1985) cites estimates of 100,000 deaths from occupational illnesses and 390,000 cases of occupational illnesses each year in the United States.

Occupational safety and health would seem to be an issue on which labor, management, and government would want to cooperate. The costs provide an economic incentive for management to limit injuries and illnesses; workers' interests are obvious since they are the ones who get sick or injured; and government has long recognized its responsibility to protect workers. Nevertheless, occupational safety and health has been the focus of continuing conflict in the United States.

Many critics have accused U.S. business and government of lacking sufficient concern about safety and health (Berman 1978; Gersuny 1981; Lofgren 1989; Nelkin and Brown 1984; Page and O'Brien 1973; Rosner and Markowitz 1987; Simon 1983; and Wallick 1972). Others have raised questions about the costs, effectiveness, and logic of the regulatory approach and of required business practices (Arthur Andersen and Company 1979; *Business Week* 1980; Viscusi 1983; Weidenbaum 1981). Managers frequently complain about the lack of cooperation of labor on safety and health issues and about workers not abiding by work rules. Occupational safety and health issues have also been important in labor-management controversies, including work stoppages and union organizing campaigns.

Occupational safety and health issues have been the focus of political debate in the United States and have often worked their way into electoral politics, most notably during the 1976 and 1980 presidential campaigns. Since its inception in 1972, the Occupational Safety and Health Administration (OSHA) has been one of the most controversial and criticized federal agencies. But even more damaging than the criticism, which it has received from virtually all corners, is the ridicule to which the agency has been subjected over the years. Practically every adult in the United States has heard jokes about OSHA "nuisance regulations" or statements of the obvious (e.g., directions on how to climb a ladder). Typical of this spirit of contempt is a poster of "The Cowboy after OSHA," which John Mendeloff described as "wearing goggles and a helmet; a net surrounds his horse, and for good measure a box is located beneath the horse's rear to serve as an 'antipollution device' " (1979:3).

The costs of OSHA regulations, as well as questions about their effectiveness, are undoubtedly at the root of much of this criticism. Lawrence S. Bacow (1980) estimated that the capital costs of OSHA compliance exceeded $3 billion annually in the 1970s. Related expenses such as general administration, training in the regulations, and materials handling and storage were also significant (Arthur Andersen and Company 1979). Various studies of the impact of OSHA have left many skeptical of the benefits derived from these costs (see Bacow 1980 and McCaffrey, Andersen, McCold, and Kim 1985 for reviews of such studies).

Occupational safety and health have become even more controversial in the aftermath of the Union Carbide tragedy in Bhopal and its serious, although less deadly, gas leak in Institute, West Virginia. Moreover, in 1986 and 1987 OSHA fined several firms, including Chrysler, Ford, and General Motors, hundreds of thousands of dollars for allegedly underrecording work injuries and illnesses. These fines have both reflected and contributed to the environment of contentiousness surrounding health and safety issues.

Labor and government have also contributed to the problems of regulating and managing occupational safety and health in the United States. In one case of underrecording, injured or ill workers and the local United Auto Workers (UAW) union apparently acted in complicity with management to allow the injured or ill workers to be laid off so that they could collect unemployment compensation from the government and under-the-table sickness and injury benefits from the company (Pummer 1986). Management's motivation was presumably twofold: first, to save money by not having to use funds set aside for workers' compensation self-insurance and second, to record low injury and illness rates and thus not be targeted for inspections.

Testimony in congressional hearings in 1988 (*Roanoke Times and World News* 1988) seemed to confirm what some have feared about U.S. government involvement in safety and health regulation, that OSHA has adjusted its figures in an effort to create a public image of effectiveness. Field inspectors testified that they were instructed to "generate numbers" even if it meant counting "scam inspections."

Occupational safety and health issues provide ample opportunities for both cooperation and conflict, making the study of these issues a useful context in which to gain insight into a wide range of management issues and the relationships among business, government, labor, and other parties. By examining these issues in both the United States and Japan, one can come to understand the similarities and differences between U.S. and Japanese management practices and institutional relationships. One can also gain insight into the emerging fields of international management and comparative social regulation.

The regulation of occupational safety and health in the United States and Japan is an important and provocative issue that has scarcely been examined. Cooperative relations between industry and government in general have been widely credited with contributing to the tremendous performance of the Japanese economy (see Halberstam 1986; McMillan 1985; Ouchi 1984; Prestowitz 1989; and Vogel 1985, for example). The role of government in promoting exports, protecting domestic industries, promoting research and development in key industries, easing the transition in "sunset industries," and providing a nonburdensome regulatory environment are well known and accurately captured by the term "Japan, Incorporated." One of the few published accounts of occupational safety and health regulation in Japan was written by Joseph L. Badaracco, Jr. (1985), who considers the issue in the context of comparing the regulation of vinyl chloride in five countries. He describes a regulatory system that relies heavily on government guidance and assistance to industry and industry cooperation with government to achieve its desired ends. With its emphasis on "administrative guidance" rather than punitive enforcement practices, safety and health regulation does indeed take on a very different form in Japan than in the United States. Business-government relations in the occupational safety and health arena adhere quite closely to the general Japan, Inc., model and provide a stark contrast to the confrontational approach to safety and health regulation in the United States.

That the styles of worker protection are different in Japan and the United States is patently clear; what is less clear is which, if either, is better. On many dimensions, Japanese safety and health practices are far superior to those in the United States, but in many other respects, they lag far behind. In still other cases Japanese practices are simply different, reflecting values Americans would not consider appropriate. One difference in the styles of worker protection is illustrated in an apocryphal story told by an AFL-CIO official involved in international occupational safety and health issues (Samuels 1986). According to the official, when there is an oil spill in a U.S. factory, a cage is built around the spill, whereas in Japan, a sign is posted, warning the workers of the hazard. One point of the story from Samuels's perspective is that in neither case is the spill cleaned up. Perhaps more accurately, the story points to the engineering orientation in the United States, that is, the use of barriers, devices, and machines to prevent injuries and illnesses, in contrast to the behavioral orientation in Japan, which emphasizes modifying the workers' behavior. As we shall see, because virtually all work involves an interaction of individuals and machines, both behavioral and engineering approaches to occupational safety and health are important.

The benefits of addressing workplace safety and health issues more constructively go far beyond the significant potential savings of human lives and the direct economic benefits of reducing the number of accidents, injuries, and illnesses. Workplace safety and health is a symbolic issue as well. It reflects management's concern for workers beyond their simple function as components in the production process. This concern can have substantial implications regarding the overall status of industrial relations in plants and companies. As a result, Japanese companies consider occupational safety and health especially important when they are beginning industrial operations in the United States. In Japan, the management of workplace safety and health is inextricably tied to and, in many ways, supportive with the manufacturing and management systems. At the national levels, the handling of workplace safety and health issues is indicative of overall labor-management-government relations. Positions taken on issues and the outcome of controversies can determine the political orientation and clout of various factions within the national body politic.

International and Political Implications of Occupational Safety and Health

The management and regulation of occupational safety and health in Japan are also important from an international perspective because of the continuing controversy over the country's trading policies. The success of Japanese firms in international markets has led many in the West to question whether that success has come at the expense of Japanese workers. The Japanese worker has been characterized as laboring for very long hours under difficult and hazardous conditions, especially during the era of rapid growth in the 1960s and early 1970s (Halberstam 1986; Kamata 1982; Sōhyō 1985). There is no doubt that the Japanese still work more hours than Americans, about 236 more hours per year on average according to an International Labor Organization study (Brauchli 1988). Statistics also indicate that occupational injury and illness rates were far higher in the 1960s and 1970s than they are today (Japan Industrial Safety and Health Association 1986).

The perception in the West that Japanese workers are abused and overworked has provided ammunition for various industries, including the automotive, that have been threatened by what was regarded as unfair competition from Japan, to demand trade protection. Organized labor in the West, primarily out of a sense of self-interest and a generalized sense of labor solidarity, has also appealed for improved working conditions and shorter workweeks for Japanese workers. Because of these experiences there is a great deal of sensitivity in Japan concerning occupational safety and health.

My own research (Wokutch 1990a) and casual observations provide evidence that the Japanese are not the only ones who are sensitive about occupational safety and health issues. In many developing countries and in countries with one-party political systems, occupational injury and illness statistics are treated as state secrets. For example, National Public Radio reported in 1987 on the capture and torture of a Korean dissident who had masqueraded as a worker in an industrial plant in order to investigate occupational hazards. Fortunately, the Japanese with whom I dealt were not as secretive.

The controversial nature of occupational safety and health creates research difficulties in any country. Japan presents unique challenges. Many visitors have noted that the Japanese have a love-hate relationship with Westerners that is complicated by a peculiar mix of feelings of inferiority and superiority. The Japanese have a sense of cultural superiority evidenced in their frequent use of the mildly derogatory term *gaijin*, which literally translates as "foreigner" or "outside person." Even individuals of Japanese descent who have been raised abroad and native Japanese who have spent extended periods of time overseas are looked on with suspicion. At the same time, individual Japanese will go to great lengths to emulate Westerners. Japan's seemingly fanatic devotion to quality came about both because of the prodding of American consultants and the Japanese desire to counteract the ridicule associated with the phrase "Made in Japan" in the early postwar years. Western fashions, customs, and entertainment are very popular in Japan, and many fashion models and actors and actresses in commercials are Americans or Western Europeans. Some Japanese even go to the extreme of having cosmetic surgery to conform more closely to Western standards of beauty. The Japanese are also dedicated, if not particularly successful, students of English. As one American living in Japan noted wryly, the Japanese seem to take English classes the way Californians go to psychiatrists; it's the "in" thing to do, and even if they don't get any better, they feel good about themselves for trying.

As a result of their conflicting feelings, the Japanese try to ensure that visitors come away with favorable impressions of Japan. This effort is particularly evident with respect to occupational safety and health.

The researcher's task is further complicated by Japanese traditions regarding candor or, more precisely, the lack of it. To understand the Japanese concept of candor, one must first understand the concepts of *tatemae* and *honne*. *Tatemae* is a professed intention or belief. We in the United States might refer to it as a socially acceptable response or the "party line." *Honne* is the real intention or practice. The Japanese refer to it as showing one's true colors. *Honne* communication usually occurs only when Japanese speak with trusted friends, colleagues, or family, or when they inadvertently let down their guard. It is unusual to find a Japanese who will discuss occupational safety and health except in terms of *tatemae*, especially with a foreigner. Most Japanese are very concerned that a foreigner see Japan as they see it or as they want it to be seen. Managers and government regulators who are willing to discuss occupational safety and health in *honne* terms are particularly rare.

The Automotive Industry

The automotive industry furnishes an important context in which to study occupational safety and health in the United States and Japan in that developments in the industry demonstrate overall trends in the U.S. and Japanese economies. The rise, fall, and subsequent partial recovery of the U.S. auto industry and the ascension of the Japanese auto industry have been used as metaphors for developments in the broader U.S. and Japanese economies (Halberstam 1986; Yates 1983). In their book *The Machine That Changed the World*, James P. Womack, Daniel T. Jones, and Daniel Roos (1990:11) quote Peter Drucker (1946), who more than forty years earlier labeled the automotive industry the "industry of industries," and they note that automobile production is still the largest manufacturing activity in the world. Lee Iacocca became one of the most recognized and respected figures in the United States by emphasizing the need for American economic renewal such as occurred at Chrysler; he has also become an outspoken critic of what he considers unfair Japanese competition. During the depths of the industry's depression in the early 1980s (and it was indeed a depression for the auto industry even if the rest of the U.S. economy experienced only a recession), the case was made for cooperation among industry, labor, and government as it has never been made before.

In 1960, U.S. manufacturers made 48 percent of the world's 16 million cars and trucks, and Japanese manufacturers made 3 percent. By 1982, the United States held only 19 percent of the world market, and Japan held 29 percent (Automotive News 1990). Studies in the early 1980s explained Japan's success as deriving in large part from greater productivity and lower costs. Labor hours required to produce comparable vehicles were estimated to be about 40 to 50 percent lower in Japanese firms than in U.S. firms (National Research Council 1982), and the cost advantage of building a typical small car in Japan was estimated to be between $1,400 and $2,000 per car (National Research Council 1982; Altshuler et al. 1984). Even adding approximately $400 for transportation expenses to the United States, Japanese producers still had a considerable competitive advantage (National Research Council 1982).

By the end of the 1980s, the Japanese had approximately the same edge in labor hours (Womack, Jones, and Roos 1990), but because of the rise in the value of the yen, which resulted from the success of the Japanese auto industry along with that of other industries, they had lost much of the cost advantage. This loss, along with political pressures, prompted Japanese producers to establish operations in the United States. With this new foreign-owned production, the United States was closing in on Japan for the distinction of being the number one auto-producing country in the world. This could scarcely be considered a victory, however, since U.S.-owned auto firms had already lost a significant share of the domestic market.

In addition to foreign competition, the U.S. auto industry has had to confront a variety of internal problems and external threats. Beginning with the publication in 1965 of Ralph Nader's book, *Unsafe at Any Speed*, the industry has been a favorite target of social critics and corporate activists. Consumer advocates, labor unions, minority groups, stockholder activists, environmentalists, and government agencies have pressed various—and often conflicting—demands on the "Big Three" (General Motors, Ford, and Chrysler). The twin "oil shocks" of 1973 and 1979 devastated the industry and threatened the survival of at least Chrysler and probably Ford as well. The Big Three made massive investments and fierce cost-cutting moves, including laying off hundreds of thousands of workers, in an effort to overhaul the design of vehicles and streamline production processes.

These competitive pressures also brought about profound changes in the nature of labor-management-government relations in the industry. Among the more noteworthy developments were the appointment of UAW presidents Douglas Fraser and Owen Bieber to Chrysler's board of directors; government-guaranteed loans to Chrysler; wage and work-rule concessions by workers; profit-sharing plans; employee involvement programs; streamlined managerial staffs; the adoption of Japanese-style, team-oriented production practices; government-negotiated Japanese import restrictions; and efforts by the Reagan administration to lower the cost of government regulations. Although the merits of any one of these changes is debatable, taken together they represent progress toward the development of labor-management-government relations that are more conducive to successful international competition.

Corporate investments, cost-cutting measures, and other institutional changes enabled the U.S. auto industry to recover in the 1980s, with Ford and Chrysler leading the way. Profitability was restored, albeit at lower levels of production and with far fewer employees than before 1979. Ominously, however, as the economy sank into a recession in 1990 and 1991, U.S. auto manufacturers once again reported huge losses.

Compounding the problems, a new threat has emerged in recent years: domestic competition from Japanese firms operating in the United States (Treece 1988). As of 1989 these Japanese transplants (including U.S.-Japanese joint ventures) accounted for 16.6 percent of automobile production in the United States (Automotive News 1990), and that figure is expected to rise substantially in coming years. As with threats in the past, Detroit had a hand in creating this one. The so-called voluntary restraint agreements, which limited the number of cars Japanese automakers could export to the United States, were negotiated largely in response to pleas for help from U.S. auto producers and organized labor. Cajoling by labor and politicians and threats of further restrictions convinced Japanese manufacturers that the only way they could increase or even preserve their U.S. market shares was to produce cars in the United States. The rise in the value of the yen relative to the dollar made this decision that much more attractive. Investment costs denominated in dollars decreased, and U.S. labor became more of a bargain.

For the most part, these transplant ventures have been highly successful. They have achieved levels of quality and productivity comparable to those of Japanese facilities[2] (Bernstein et al. 1986; Brown and Reich 1989), and they have been able to retain an estimated $600 per vehicle cost advantage over their U.S. competitors (Ingrassia 1990). One expected milestone in the success of the Japanese transplants is that Toyota or Honda will most likely pass Chrysler in U.S. market share in the next few years.

Corporate Social Responsibility

Over the years, considerable attention has been focused in management literature, executives' speeches, the popular media, and demands by interest groups on the social responsibility of the corporation.[3] But although there are clearly implications regarding corporate social responsibility (CSR) in the great majority of writings about Japanese

Figure 1-1. Corporation-Stakeholder Relations

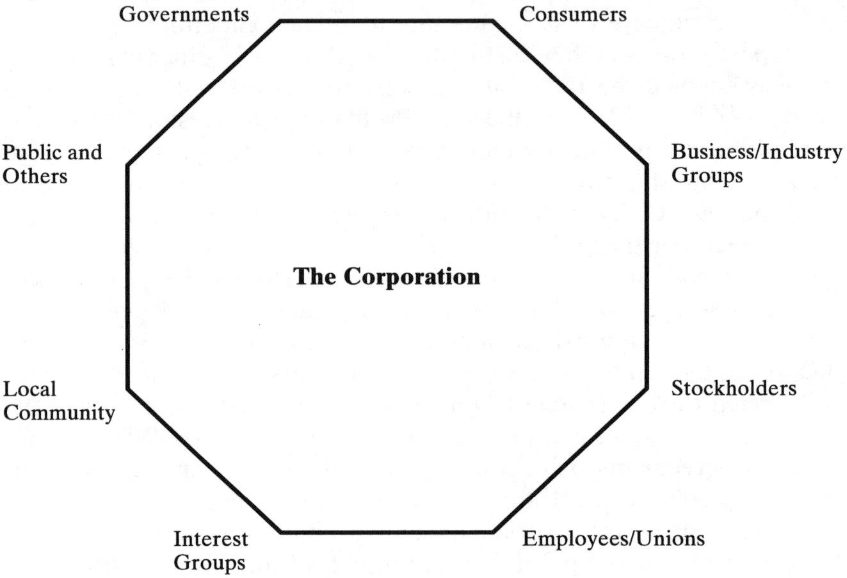

management, surprisingly little explicit attention has been devoted to the issue as it relates to Japanese business firms.[4]

Despite arguments put forth by Milton Friedman (1970) and Theodore Levitt (1958), a virtual consensus has emerged that business should assume certain social responsibilities for both ethical and profit-related reasons. Notable landmarks in the development of this consensus are statements by the Committee for Economic Development (1971) and the Business Roundtable (1981). There is now much less debate about whether firms have any social responsibilities other than maximizing profits; it is generally taken as a given that they do. Rather, attention is focused on what these responsibilities are and how they should be managed.

Figure 1-1 illustrates a common conceptualization of CSR in the United States. Corporate management is said to be responsible for balancing the interests of the various stakeholders of the corporation (employees, stockholders, customers, suppliers, government, the local community, the general public, and various interest groups) in an effort to promote the common or social good. No prioritizing of these stakeholders is usually attempted, and the relative importance of these groups would differ from issue to issue.

The reasons suggested for socially responsible behavior can be summarized in two basic motivations: enlightened self-interest and ethical or altruistic considerations. Enlightened self-interest (ESI) is the notion that by undertaking certain actions that are beneficial to various stakeholder groups or to society as a whole, a corporation will prosper in the long run, even if it has to bear short-run costs. The idea is that one can do well financially by doing good morally. Even the very restricted view of CSR Friedman proposed would not conflict with the notion of enlightened self-interest, although he does not use this terminology. Friedman has argued that "there is one and only one social responsibility of business—to use its resources and engage in activities designed to increase its profits so long as it stays within the rules of the game, which is to say, engages in open and free competition without deception or fraud" (1970:126). According to this view, actions motivated by self-interest, whether "enlightened" or not, and that enhance long-run profitability are socially responsible. Indeed, Friedman even considers greed socially desirable because of its economic benefits.

There are also those who propose essentially ethical or altruistic arguments for CSR (Donaldson 1982; Frederick 1986; Freeman and Gilbert 1988). They argue that ethical considerations should in some circumstances supersede financial considerations in dictating corporate activities. Thus, corporations should undertake certain activities to benefit society or stakeholders of the organization regardless of the effect, in the short or long run, on the bottom line. Ethical arguments for CSR activities derive from the view that, just as individuals have certain moral responsibilities, so too does a corporation. A corporation can, in fact, be viewed as an artificial person with the ability to do great harm or good. Moral duties normally ascribed to individuals, such as the duty not to harm others, to keep promises, and to help others if one can provide great benefit at relatively little cost to oneself, would be said to apply as well to a corporation by advocates of this view.

Altruism is a more controversial motivation for CSR than enlightened self-interest. There are those who defend the use of corporate resources for altruistic purposes, while critics such as Friedman condemn the practice. Notable among the staunchest defenders are corporate executives themselves, who frequently justify CSR activities on the grounds that they are simply the right thing to do. In a 1976 survey (Holmes 1976), 54 percent of the chief executive officers of 192 U.S. companies said they would engage in CSR activities even if there was no probability of short-term or long-term profits.

Given human nature, in most cases, enlightened self-interest is likely to be the dominant motivation for CSR. In Japan its predominance over ethical considerations is particularly obvious. I will return to this point in later chapters.

Occupational safety and health clearly affects an important constituent group—employees—and firms typically are judged to have social responsibilities concerning its management. The various benefits that accrue to a firm in both the short and long run from occupational safety and health activities justify these activities as being in the firm's enlightened self-interest. For example, certain expenditures to reduce the hazards of production processes or increase worker awareness of safety and health could produce short-term financial benefits for the firm, such as decreased accident- and illness-related costs from medical expenses, absenteeism, short-term compensation to injured and ill workers, and equipment damage and lost production time. And, as recent events have shown, expenditures can help prevent firms from having to pay hefty fines to OSHA.

A firm also derives long-term benefits from safety and health activities. These include avoiding the costs associated with occupational illnesses that take years to develop and long-term increases in workers' compensation contributions. Safe and healthful working conditions may improve the morale of workers and lead to greater productivity. In addition, unions such as the UAW have focused on allegedly unsafe and unhealthful working conditions during drives to organize nonunionized plants (including Japanese-owned operations). A firm hoping to avoid unionization for financial and other reasons may find that it is in its long-term interest to provide a margin of safety and health over and above that required by law. Long-term public relations benefits may also result from a firm's health and safety activities. The firm may find it easier to attract qualified workers if it has a reputation for being concerned about its employees, and consumers may be influenced to purchase or not to purchase products based on the firm's reputation as an employer.

The aphorism "Safety pays," frequently used by labor, management, and government officials involved in safety and health activities, makes this economic argument succinctly. It suggests that activities that promote safety (and, it is generally understood, health) are in a firm's enlightened self-interest and will maximize the firm's profits in the long run.

Implicit in the meaning of corporate social responsibility suggested by figure 1-1 is the possibility that firms will undertake activities that benefit society or some corporate stakeholders but that are detrimental to the finances of the firm, even in the long run. Such activities are described as altruistic. That is, they promote the interests of others (society, employees, consumers, and so on) without any profit incentive or expectation of profit.[5] A firm's occupational safety and health activities would be considered altruistic if it was expected that the cost of some voluntarily undertaken safety- or health-related practice would exceed all short-term and long-term benefits of that practice.

These concepts relate to two important questions regarding occupational safety and health: (1) are firms doing everything they can with respect to occupational safety and health that is in their own enlightened self-interest (or long-term profit interests), and (2) are firms undertaking some activities in this area for altruistic purposes? While the long-term profit implications of activities that are undertaken or of potential activities that are not undertaken are extremely difficult or impossible to ascertain precisely, it is possible to make reasoned judgments on these matters. In doing so, I will address both of these questions.

Cooperation and Conflict

Because the cooperative and conflictual aspects of labor-management-government relations pertaining to occupational safety and health are so important, it is first necessary to define the concepts of cooperation and conflict. Although these terms are commonly used in everyday language, there is considerable scholarly debate about their precise meanings. A general understanding of these terms will suffice for our purposes here.

Conflict may be viewed in situational terms as a process in which one party perceives that his or her concerns have been or are about to be frustrated by another party (Thomas 1976). In behavioral terms, it is behavior in which one party seeks the advancement of its own interests in its relationships with others (often at the expense of the others' interests). Cooperation or a cooperative situation will be taken to mean one in which the goals of the parties are linked in such a way that a party can attain his or her goals only if the others involved can attain their goals (Deutsch 1973:20).

The everyday sense of conflict and cooperation is that they are polar opposites, with conflict viewed as bad and cooperation viewed as good. Conflict is thought to involve such negative features as the lack of recognition of shared goals and, as a result, the inability to achieve them, the resentment of the losing party in situations that are reduced to win-lose terms, and potential disruptive and destructive relations, producing emotional or even physical harm. Cooperation is believed to enhance a group's or an organization's ability to achieve a common or shared goal while at the same time improving the quality of personal relationships within the group.

Scholars have noted that there are positive features of conflict and negative features of cooperation. Morton Deutsch (1973) argues that conflict tends to stabilize and integrate relationships, that it can revitalize existing norms and establish new ones in the relevant group or organization, and that it can help define the leadership in organizations. Other benefits include increased motivation, interest, and creativity; higher-quality ideas and solutions; and, in the case of external conflict, increased internal cohesiveness (Deutsch 1973; Thomas 1976).

Deutsch (1973) further argues that cooperation can lead to overspecialization, provincialism, and the inability to see the "whole picture." "Group-think" decision making that does not challenge crucial but questionable assumptions can result in nepotism, strained relationships with others outside the group, and, ultimately, lower levels of goal attainment.

Others dispute the unidimensionality of cooperation and conflict. Thomas L. Ruble and Kenneth W. Thomas (1976) and Thomas (1976) propose instead a two-dimensional model of conflict behavior (see fig. 1-2). This figure refers to the behavior of one party toward another party in a conflict situation. Assertiveness (i.e., emphasis on satisfying one's own concerns) is plotted on the y-axis; cooperativeness (i.e., emphasis on satisfying others' concerns) is plotted on the x-axis. Ruble and Thomas (1976) label five regions on the figure as five types of behaviors: competitive (high assertiveness, low cooperativeness); avoidant (low assertiveness, low cooperativeness); sharing (intermediate assertiveness and intermediate cooperativeness); accommodative (low assertiveness, high cooperativeness); and collaborative (high assertiveness, high cooperativeness). Competitive behavior as defined by Ruble and Thomas corresponds to the behavioral view of conflict specified earlier. The term "confrontational" is also used in this sense in this book.

Conflict behaviors of the type characterized by each of the five regions are appropriate for different situations. Thomas contends that collaborating, for example, is the most appropriate behavior in the following situations:

1. To find an integrative solution when both sets of concerns are too important to be compromised.
2. When your objective is to learn.
3. To merge insights from people with different perspectives.
4. To gain commitment by incorporating concerns into a consensus.
5. To work through feelings which have interfered with a relationship. (1979:161)

Figure 1-2. Conflict-Coping Behavior

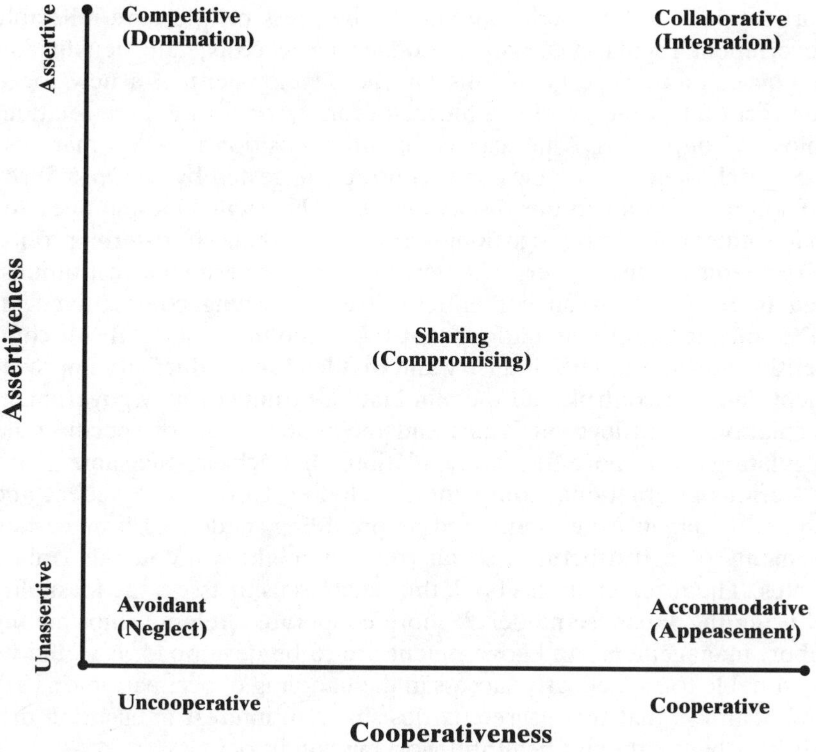

Source: Adapted from Ruble and Thomas 1976:145.

As we shall see, these conditions are common in conflicts pertaining to occupational safety and health.

These notions of conflict and cooperation, along with avoidance, accommodation, compromise, competition, and collaboration, will be referred to in subsequent chapters to explain labor-management-government relations pertaining to occupational safety and health. Figure 1-2 will be particularly useful for describing and comparing relationships in the United States and Japan.

New Social Contract

In light of America's weak international trade situation in recent years, there have been numerous reexaminations of how U.S. economic and institutional relationships are structured. A virtual consensus has emerged that many of the nation's economic problems are caused by the confrontational relationships among business, government, and labor; the lack of long-term thinking by business people; and inflexible government regulations, which produce more costs than benefits. In response, there have been calls for the development of a new social contract that would provide a blueprint for restructuring these relationships and improving America's competitive position in world markets.

Several aspects of a new social contract suggested by *Business Week* (1980) are relevant to our discussion here. These include the need for management to emphasize long-term rather than short-term profits; agreements among leaders of labor, business, government institutions, and interest groups on contentious issues affecting competitiveness; labor-management negotiations that take account of the difficult competitive position of U.S. industry and that lead to productivity improvement and cost control; and the elimination of unnecessary government regulation of business activities and the replacement of specific rigid regulations with more flexible regulations that achieve the same goal.

Various international competitors including Germany, Sweden, and especially Japan have been cited as providing models of how certain elements of a restructured social contract might work in the United States. The research in this book thus enables us to assess the feasibility of using the Japanese model. If more cooperative relationships among labor, management, and government are to be developed, it would be reasonable to expect early success in the handling of occupational safety and health in that these three parties share an interest in ensuring that workers have safe and healthful working conditions.

The U.S. operations of Japanese auto producers provide natural laboratories in which to examine attempts to adapt elements of the reportedly more cooperative Japanese social contract to the U.S. work environment. Moreover, the working conditions and labor-management relations at the transplants have been the subject of both high praise and damning criticism (Brown and Reich 1989; Fucini and Fucini 1990; Holusha 1989; Kendall 1987; Kraar 1989; Parker and Slaughter 1988). Criticism was especially severe during the unsuccessful 1989 UAW organizing drive at Nissan, which focused in large part on allegedly inferior safety and health conditions there. It is hoped that this book will shed some light on these controversies and help counter simplistic stereotypes.

Comparative Regulatory Policy

Although this research seeks to explore a number of issues that have not been considered previously, it is important to relate it to several streams of literature to which it is relevant. Steven Kelman's book (1981) on the regulation of occupational safety and health in the United States and Sweden is the classic in the field of comparative safety and health regulatory policy. Although the Swedish and Japanese occupational safety and health systems are very different in many important respects, a number of Kelman's findings about the Swedish occupational safety and health regulatory system relative to that in the United States are relevant to Japan–United States comparisons. Kelman observed that the Swedish system is more flexible, cooperative, and guidance-oriented and that the American system is more rigid, adversarial, punitive, and legalistic. The same could be said of the Japanese occupational safety and health system compared with that of the United States.

The existence of these characteristics of the American regulatory system vis-à-vis other regulatory systems is a theme that has been repeated in several studies of regulation in the United States, Europe, and Japan. These include Badaracco's (1985) comparison of vinyl chloride regulation in five countries, including the United States and Japan; Ronald Brickman, Sheila Jasanoff, and Thomas Ilgen's (1985) study of chemical regulation in four European countries and the United States; David Vogel's (1986) study of environmental protection policies in the United States and Britain; Graham K. Wilson's (1990) study of occupational safety and health in the United States and Britain; and my study (1990a) of occupational safety and health in five industrialized countries, including the United States and Japan, and in one developing country, Kenya.[6] The disadvantages of the U.S. regulatory system have been consistently lamented in this literature. But certain advantages of the more adversarial approach have also been identified, as have disadvantages of the more cooperative approach (Kelman 1981; Vogel 1986; Wokutch 1990a).

Two studies cited above (Fucini and Fucini 1990 and Kendall 1987) consider the management of safety and health in Japanese automotive transplants operating in the United States. Neither, however, provides a scholarly treatment of the subject. The first, based on extensive interviews, is an in-depth description by two journalists of working life at Mazda's U.S. auto plant. It includes one chapter that is sharply critical of the plant's safety and health practices. The second study is a brief practitioner article in a trade journal that paints a highly flattering portrait of safety and health management practices at U.S. Nissan and Honda plants.

This study provides a perspective that was not available in most earlier studies: that is, the investigation of the management of occupational safety and health at the company and plant level in conjunction with an examination of national (and, to a limited extent, regional) regulatory policies and practices. Thus, this study enables the reader to compare regulatory policies and assess how they affect individual companies, plants, and workers. It also enables the reader to see how well management practices regarding safety and health can be transferred across national boundaries to a different regulatory environment.

Methodology

This research employed several methodological approaches, enabling me to compensate for the inherent limitations likely to be associated with any one approach (Webb et al. 1966). The principal approach used was the directed interview, but documentary information, safety and health and regulatory data, observation of work environments and work practices, and numerous informal conversations were also important. While much of the research in Japan was conducted in four months under a Fulbright research fellowship, subsequent communications enabled me to clarify issues and access data over approximately five years.

I interviewed scores of industry, labor, and government officials and auto industry workers in both the United States and Japan during the course of my research. A total of twelve plants (eight in Japan and four in the United States) of six major Japanese auto firms were studied. Also studied were plants of several Japanese subcontractor firms. During an earlier phase of this research, I examined a similar number of plants, both U.S.-owned and foreign-owned and operating in both the United States and other countries, including Germany, Sweden, Great Britain, and Kenya (Wokutch 1990a). Of the plants considered in this study, the greatest attention was focused on the new U.S. factory and the main Japanese factory of one firm that I identify here as Jidosha, from the Japanese word for automobile, *jidōsha*.

Auto industry officials interviewed included corporate-level and plant-level managers, safety and health and medical staff, production workers, foremen, and other supervisory personnel. I also interviewed union officials concerned with safety and health at the plant level, company level, and national level in Japan and the United States. In addition, I spent considerable time with company officials and workers in less formal situations in which actions were less guarded and comments more candid. Most of the officials I interviewed were from the auto industry, but I also consulted union officials from other industries.

Access to corporate and union officials at Jidosha was exceptional. I had the opportunity to discuss safety and health issues with most top-level officials, including the chief executive officer of Jidosha and the president of the Jidosha union, as well as chief operating officials concerned with safety and health.

Government officials from OSHA and its Japanese counterpart, the Labor Standards Bureau (at both national and regional levels) were interviewed. Officials from one state-level occupational safety and health regulatory agency having jurisdiction over Jidosha's operations in the United States were also consulted, as were numerous officials from other related organizations, both public and private.

I spent considerable time simply observing work at Jidosha, at Jidosha subcontractors, and at other major Japanese auto firms. Although I was accompanied in most cases by management personnel whose presence made workers guarded in their behavior and actions, I also traveled through Jidosha facilities alone. Dressed in a Jidosha uniform, I was indistinguishable at a distance from Japanese workers. I also had several opportunities to speak with workers alone or with only an interpreter. There is thus good reason to have confidence in the authenticity of the observations and interview data on which I report.

This book considers both the regulation and management of occupational safety and health because these are highly interrelated. It also presents research entailing several levels of analysis. That is, I consider research on worker safety and health issues at the national, industry, firm, and plant levels. This approach makes possible a fuller understanding of these issues and accounts for the limitations of analysis on only one level.

Organization and Focus

This book examines the national style of worker protection in Japan and compares it with that in the United States. The emphasis is on the Japanese style of worker protection, however, since this research provides the greatest contribution to the literature. The literature on occupational safety and health in the United States is plentiful; English-language literature on occupational safety and health in Japan is virtually nonexistent. A more detailed analysis of the occupational safety and health system in the United States is contained in Wokutch 1990a. The interested reader might also want to consult other sources, such as Ashford 1976, Berman 1978, Gersuny 1981, Lofgren 1989, McCaffrey 1982, Mendeloff 1979, Nelkin and Brown 1984, Simon 1983, U.S. Congress Office of Technology Assessment 1985, and Viscusi 1983.

Chapter 2 provides general background on the environmental context in which occupational safety and health systems function in Japan, including organizational aspects and the manufacturing management system; historical considerations; demographic, geographic, and economic considerations; labor-management-government relations; and political and government considerations. The motivations and determinants of corporate social responsibility in Japan relative to the United States are also considered.

Chapter 3 provides a description of the Japanese occupational safety and health system, including existing legislation and practices. It also contains a comparison of occupational safety and health regulatory activities conducted by OSHA in the United States and by the Labor Standards Bureau in Japan. It focuses particularly on staffing levels as well as workplace inspection and enforcement experience.

Chapter 4 shifts the level of analysis from the national level to the company and plant levels. This chapter is an in-depth description of the operation of the occupational safety and health system at Jidosha. It focuses primarily on the operation of Jidosha in Japan (in particular at its flagship plant), but it refers occasionally to its operations in other countries.

Chapter 5 focuses on the operation of Jidosha in the United States at its recently established auto manufacturing and assembly plant. Comparisons of the operation of Jidosha safety and health systems in Japan and in the United States and injury and illness statistics are also considered.

Chapter 6 focuses on work injury and illness experiences in the United States and Japan at national, industrial, firm, and plant levels. In conjunction with data from Jidosha, work injury and illness data from the U.S. and Japanese operations of two Japanese automotive firms, fictitiously named Kuruma and Norimono, are reviewed. Special attention is given to the difficulties involved in comparing injury and illness experiences cross-nationally.

Chapter 7 summarizes the research findings and provides conclusions and recommendations. These recommendations include discussion of the feasibility of the United States or Japan adapting the best of the other's national and corporate occupational safety and health practices and suggestions for how this might be done. The implications of this for cross-national learning about aspects of each other's institutional relationships, management practices, and regulatory approaches are also considered.

2

The Environmental Context for Occupational Safety and Health Activities in Japan

To understand the regulation and management of occupational safety and health in Japan, one must first understand the environment or context in which occupational safety and health activities occur. As we shall see, this context defines the limits of managerial and regulatory prerogatives regarding workplace safety and health. In this chapter I describe this context so that the reader may understand and appreciate subsequent analyses of safety and health at national, industrial, firm, and plant levels. Those aspects of the environment that I consider are organizational conditions and the manufacturing management system; historical conditions; demographic, geographic, and economic conditions; labor-management-government relations; and political and government conditions. In addition, so as to provide a broader context for understanding the Japanese style of worker protection, I then consider the relationship of these environmental conditions to the motivations and determinants of CSR in Japan.

Environmental Conditions

Organizational Conditions and the Manufacturing Management System

The way in which the management of occupational safety and health fits into the overall scheme of Japanese management can be seen in the relationship of safety and health management practices to better known aspects of the Japanese manufacturing system, including productivity and quality-improvement activities and inventory control. Some of these innovations of the production management system have been cited as crucial to Japan's industrial success. I will briefly review the more distinctive features of the production system so that I can explore the relationship of safety and health management to it.

The typical U.S. system of mass production was originated by Henry Ford. It consists of large-lot production for one model of automobiles, refrigerators, television sets, whatever, and relatively long set-up times for switching production to other models. Intermediate parts are stockpiled, resulting in large inventories of raw materials and works in process. These stockpiles serve as buffers between work stations and workers so that they can be relatively independent of one another. They can continue working even if there is a problem elsewhere on the line. Quality inspections typically take place at the end of the production cycle, and defective products are either repaired or discarded.

In contrast, Japanese production is characterized by flexible manufacturing systems, small-lot production runs for parts, a short set-up time for switching to production of different models, an interdependence of workers and work stations in the assembly system, small inventories, and workers taking responsibility for quality on the production line. The management philosophy and the manufacturing system are designed to minimize waste and inefficiency and to maximize efficiency and the harmony of interdependent operations. Womack, Jones, and Roos (1990) refer to this system as lean production. They suggest that in the future automotive and other manufacturing plants will have to operate under this system to survive.

The inventory control system illustrates how lean production operates. Toyota's well-known *kanban* inventory system (Monden 1983; Schonberger 1982) and similar approaches are designed to keep inventories of raw materials and semifinished parts to a minimum. In addition to substantial cost savings, this approach brings breakdowns in the production system to everyone's attention—the Japanese notion of *jidōka*—and the problem is corrected immediately. In contrast, in the traditional U.S. production system, the semifinished parts go into inventories, and a problem could persist for some time without anyone knowing about it. Even when defects are recognized in traditional U.S. plants, the thinking has often been to let quality control or repair shop personnel worry about it.

The Japanese are fond of illustrating the advantage of their production system by way of an analogy. They show a body of water with many large rocks in it. When the water level is high, the rocks are completely submerged and therefore pose unseen hazards to passing ships. When the water level is low, the hazards can be seen and precautionary measures can be taken.

The management of workplace safety and health is quite consistent with other aspects of Japanese production management. In addition to the human and physical damages that can result, accidents cause serious disruption for Japanese manufacturers because they operate with so little slack. Thus, accidents are viewed as an extreme form of inefficiency and disharmony of operations that should be eliminated. It is in this spirit, that no accidents are considered acceptable, that "zero-accident campaigns" are undertaken.

Zero-accident campaigns are patterned after the more widely known zero-defect campaigns in Japanese plants. These seek to eliminate all defects in production processes and finished products. In fact, the two campaigns are closely intertwined because of the interdependence among safety, quality, and productivity, which the Japanese are fond of emphasizing.

The relationships among safety and health, productivity improvement, and quality improvement are also seen in *kaizen* activities. These activities vary from company to company, but, in general, they are cooperative efforts to analyze the component elements of a work task and plan ways to improve how it is done both from a quality and a time perspective. The intent is said to be to make the worker's task easier to perform, not to make him or her work harder. (It is sometimes described as designed to enable the worker to work smarter, not harder.) Still, managers attempt to ensure that 100 percent of each worker's time is directed toward production or other company goals. Workers' tasks are often made easier by eliminating obstacles that not only impede work but are potentially hazardous.

There are other interesting differences between the management of safety and health and the management of quality in the auto industries of the United States and Japan. One of the reasons that the quality is higher in Japanese plants is that the workers are involved with and responsible for quality. This human involvement includes but goes beyond suggestion systems and quality circles. A concern for quality is integrated into the production system. Each worker must be concerned with the quality of his or her own work and, moreover, with that done in the previous stages of production on which he or she depends. Each worker must constantly inspect work received from the preceding stage to ensure that he or she can build on it; likewise, the worker recognizes that his or her own work is subject to inspection by the next worker on the line. If there is a problem, the worker is encouraged to call for help or even to stop the production line. This system allows for the almost immediate identification and correction of problems. Thus, it has often been said that "you can't inspect quality into a product." That is, you can't rely on inspectors at the end of the production line to ensure product quality. The concern for quality must be integrated into the production system.

An analogous situation exists with respect to safety and health. As with quality, safety and health can be promoted to a large extent through engineering controls, but machines break down, and humans can sometimes operate machines incorrectly or subvert the safety and health controls on the machines. In addition, nonroutine procedures, such as maintenance and construction, typically are not automated and require high levels of human involvement.

Safety and health experts, like quality control experts, can look for problems, but they cannot check everything. Line managers can play an even more important role in managing health and safety, but, they cannot be everywhere either; they must rely on individual employees to check for safety and health hazards both to themselves and their co-workers. Such worker involvement in safety and health matters has been developed to a high degree at Japanese auto companies.

Although they occasionally overstate the case, the Japanese explain the relationship between safety and other aspects of the production process using a number of slogans, which they repeat like mantras. One, promoted by the Labor Standards Bureau, is *anzen nakushite seisan nashi* ("There can be no production without safety"). Japanese managers and workers alike also claim that safety, production, and quality are "all part of the same equation." Another popular slogan stresses the need to avoid *muri* (excess), *muda* (waste), and *mura* (unevenness). The relationship of efficient production to safety and health is also seen in Toyota's four alliterative safety and health principles: *seiri* (getting rid of redundancy), *seiton* (putting in order), *seiketsu* (clean and hygienic), and *seisō* (cleansing). These slogans reinforce the theme that harmony, *wa*, in production, like harmony in interpersonal relationships, is highly valued and that accidents disrupt this harmony.

Likewise, Honda emphasizes the "Four Cs" of safety and the "Ten Golden Rules" of safety which reflect company and social values. The meaning of the Four Cs—control, conformity, cleaning up, and cleanliness—is, for the most part, obvious. Control, in the context in which it is used here, however, refers to distinguishing between activities that are necessary for production and those that are unnecessary and getting rid of the latter. Conformity is described as "putting necessary items in a predetermined location for easy accessibility and use." It is clear that conformity also refers to compliance with established norms of behavior. The relationship of these principles to productivity management is also obvious. In particular, the description of control noted above is consistent with Japanese inventory control practices, which have been credited with creating production efficiency in Japanese companies.

Honda's "Ten Golden Rules" of safety are explained in an English-language safety and health manual for new employees at the U.S. plant entitled *There Can Be No Production without Safety*. The rules are interesting for their emphasis on practices that are not obviously safety-related:

1. "First, Before Going Into the Plant, the Checks." This is similar to the morning safety check and call at Jidosha described in chapter 4. The manual suggests checking one's cap, uniform, safety shoes, and name badge.
2. "Clear Greetings in a Loud Voice." This rule can be understood only in the context of the behavioral approach used to promote safety and health at this and other Japanese plants.
3. "Inside the Plant." This rule is explained as follows: "Safety is your concern, and everyone's concern, so take the lead in cooperating for safety. Always start work equipped with the knowledge and skills you'll need for the day's work. Particular care is needed during on-the-job training."
4. "Observe the Rules of Safety." Emphasis is on knowing the specific steps, methods, and rules for each job and following them without fail.
5. "Control and Conformity—The Basics of Safety." The explanation repeats that of the Four Cs.
6. "Correct Wearing of Protective Equipment." This is explained by analogy to baseball, in which catchers and batters always wear appropriate safety equipment.
7. "Life by the Rules." This refers to life off the job and suggests refraining from activities such as heavy eating or drinking, which could lead to lethargy and tiredness, or worrying about home life, which could affect concentration and performance on the job and lead to injuries.
8. "Unsafe Conduct and the Resulting Catastrophe." This rule points out specific work practices, such as having bad posture, playing practical jokes, wearing improper clothing, or wearing clothing improperly, that are likely to cause problems and therefore should be avoided.
9. "Keep it Tidy for Everyone." No explanation is given. In Japan, the concern for others in one's group needs no explanation.
10. "If, Despite Everything, an Injury Occurs. . . . " This rule explains the actions that should be taken in the event of an injury. The worker is encouraged to act quickly but coolly if a colleague is injured—to report the injury and provide assistance but to avoid injury to oneself. The injured worker is told to follow the doctor's instructions.

Like the Four Cs, these Golden Rules have social and value implications that go beyond safety and health matters. They are clearly intended to promote the corporate culture and values that are deemed important throughout Japanese society.

The Japanese manufacturing system and the philosophy of management characterized by the Four Cs and the Golden Rules do enhance safety and health. The high interdependence of this manufacturing system, caused by the extremely small inventories of work in process, results in relatively higher costs in the event of accidents than is the case in large-lot production, where more individual work stations can be idled without stopping overall production. Japanese workers are aware of their responsibility to be careful and take considerable precautions to follow work rules. The flexible manufacturing system, which requires multiskilled workers who are rotating through jobs, also makes the workers' jobs more interesting, thereby making them more attentive and, one would expect, less likely to be careless and cause an accident. Worker participation in setting goals, including those related to safety and health, leads to greater worker commitment to achieving those goals.

The emphasis on workplace safety and health is also a critical component of Japanese management philosophy regarding the treatment of workers. The notion that a paternalistic corporation provides lifetime employment and a wide range of other needs in exchange for a worker's loyalty and devotion to the corporation would be ludicrous without concern for the basic needs of safety and health.

Nonetheless, there are significant conflicts between the Japanese approach to safety and health and the manufacturing system. Continuous efforts to improve productivity and quality and to use 100 percent of a worker's time impose considerable stress on individuals both physically and psychologically. Accidents can result from hurrying, for example, and injuries and illnesses are more likely to occur when workers are tired. Likewise, the minimal slack of the Japanese manufacturing system puts great pressure on workers to cover up injuries or illnesses so as not to disrupt production and bring shame on themselves and their co-workers. In addition to the obvious harm to workers, such cover-ups also conflict with the *jidōka* manufacturing concept of bringing problems to the surface where they can be immediately addressed.

Historical Conditions

The management and regulation of occupational safety and health in Japan, as in any country, must be considered in the context of national conditions and traditions. It has been argued that a significant degree of environmental determinism exists regarding occupational safety and health (Wokutch 1990a). In other words, social, economic, legal, political, and other nation-specific factors dictate to a great extent what can and cannot be done regarding occupational safety and health. Distinctive national styles of worker protection arise as a result. In this section I explore the relevant historical conditions.

Although much attention has been focused on Japan's postwar recovery, the onset of its remarkable economic progress predates World War II. Until the second half of the nineteenth century, Japan was still a feudal society. Industrialization began following Japan's forced opening to the West by U.S. Commodore Matthew C. Perry in 1854 and the subsequent Meiji Revolution in 1868. Japan was largely isolated from the rest of the world—especially from the West—throughout much of its history, in part because it is an island nation and in part by design. Recognizing the threat of colonization and fearing the contamination of its culture from missionaries and traders, Japan expelled nearly all Westerners and virtually cut off contact with the Western world in the early seventeenth century. It maintained limited contacts through Dutch traders, who were allowed only in the port town of Nagasaki. This self-imposed isolation lasted until Perry arrived with his warships and demanded that the Japanese open up trade with the West.

As a result of its isolation, Japan lagged far behind the United States and Western Europe in industrializing, which had implications for occupational safety and health. Occupational hazards were late in developing, late in being recognized, and late in being confronted by government and industry.

Many of the values and customs governing relationships among the Japanese in the feudal era survive, albeit in modified form, in modern Japan. The emperor, who is the ultimate symbol of the nation and who was widely believed to be of divine origin, was the primary focal point of loyalty for the Japanese in feudal days. Unlike the Chinese, from whom the Japanese borrowed many concepts of ethics and norms of behavior, the Japanese did not make this loyalty conditional on the benevolence or virtue of the emperor; it was absolute (Benedict 1946). At the time of the Meiji Revolution, however, the emperor was just a figurehead. The real power belonged to the shogun, who was the most powerful member of the samurai or warrior class. The samurai were employed by the nobility (daimyo) to enforce their rule in the fiefdoms assigned to them by the shogun. These daimyo were given the right to rule over their fiefdoms and to profit from the use of the land in exchange for their pledges of allegiance to the shogun. The peasants, in turn, owed this same allegiance to their lords in exchange for the significant privilege of working the scarce arable land. The network of duties and debts of gratitude requiring repayment (*giri*) that undergirded the Japanese feudal system have their counterparts in modern Japanese industrial relations. Thus, the paternalism of Japanese employers and the loyalty of their employees can be understood in this historical context.

Traditions regarding the cultivation of rice played an early role in developing the strong sense of group consciousness that dominates interpersonal relationships in Japan today. Certain economies of scale and the need for major tasks such as planting and harvesting to be completed within relatively short time spans necessitated cooperation among villagers in the cultivation of rice. The importance of rice to the Japanese is demonstrated by the use of the same word, *gohan*, for the English words *rice* and *meal*. (The Japanese insist in trade negotiations on the maintenance of a complete ban on rice imports into Japan, even in the face of intense international pressure to open its market.)

The strong sense of affiliation and duty to one's group (however that is defined) for which the Japanese are renowned was very much in evidence in preindustrial Japan. Peasants were organized in small groups for the purposes of paying taxes to the lord, using the currency of rice. Everyone in the group was responsible for everyone else. If one person failed to pay the rice tribute, all were executed. This tradition, in more civilized fashion, survives in modern times. Thus, schoolchildren may have to make sure that the slowest learners have mastered a lesson before the class can move on to the next one. And if one child misbehaves, the entire class can be punished, thus resulting in strong peer pressure to conform to social norms. The team concept of work group organization in manufacturing plants reflects the same historical antecedents and entails the same peer pressure.

Perry's incursion into Japan revealed the weaknesses of the shogunate and its policy of isolationism. The forced internationalization brought about a number of problems for the Japanese economy with which the shogunate was ill prepared to cope. Public discontent led to the predominantly bloodless Meiji Revolution, which ended the shogunate and restored the power of the emperor at the head of a constitutional monarchy. Profound economic, social, and political changes followed.

Toshiyuki Tamura and Sohtaro Kunihisa (1982) identify four subperiods or stages in the industrialization process in Japan. During the first period, 1868–95, policies for modernization were pushed forward under the slogan *Fukoku Kyōhei* (Rich Country, Strong Military). The acquisition of foreign military technology had actually begun under the pre-Meiji government of Tokugawa Ieyasu as it tried to obtain the means to fend off the revolution as well as external military threats. The Meiji government continued this pattern, attempting to build the country both militarily and industrially (Tsurumi 1978). Fearing external military intervention, the Meiji government expanded contacts with more advanced nations to acquire basic scientific knowledge to bolster national defense.

The Ministry of Industry, Kobushō, was formally established in 1870 to serve as the government vehicle for the promotion of industrial development (Tsurumi 1978). During this period, considerable emphasis was placed on absorbing foreign technologies in an attempt to close the gap with the West. One strategy was to hire foreign engineers, scientists, and technicians to start up operations and train Japanese who would eventually replace them. Two factory systems subsequently evolved. Factories established by the government served as models of technology and organization (Patrick 1976). Aspects of the work environment tended to be western in orientation. Workers benefited from reduced working hours, Sunday holidays, and improved compensation and working conditions. By contrast, workers at private factories worked longer hours, for lower wages, under minimal conditions (Tsurumi 1978; Patrick 1976). Work-related health problems, especially in the private factories, were common. New machines, both imported and designed in Japan, led to new risks and new industrial injuries and illnesses throughout industry (Japan Institute of Labour 1982).

During the second period, 1896–1919, industrialization began to "regularize" (Tamura and Kunihisa 1982). Although government ownership of factories had declined by this time, the government's role in guiding the economy had been strengthened. In the early 1900s, the government made the first significant attempts to address the growing problems of safety and health in the factories. The Mine Law of 1905 and the Factory Law of 1911 provided for general improvements in working conditions in mining and other industries. It placed limitations on child labor and labor by women and imposed other restrictions that had safety and health implications (Japan Institute of Labour 1982). These laws are discussed in greater detail in the next chapter.

Although Japan trailed the Western industrialized nations in efforts to improve safety and health conditions, the relatively short time span by which it trailed is noteworthy given the West's large lead in industrialization. For example, Great Britain enacted its first major pieces of safety and health legislation in 1802, Germany in 1884, and Sweden in 1889 (Wokutch 1990a). The United States, which relied on state-level regulation of health and safety at work, did not have any major safety and health legislation at the national level until the passage of the Occupational Safety and Health Act in 1970.

The third period of industrialization, 1920–45, saw first an increase and then a decrease in the emphasis on improving benefits and working conditions in general and on safety and health conditions in particular. During the 1920s, employers attempted to preempt the rise of unionization by improving employee benefits (Fujita 1984). Of note, a national industrial safety week was established in 1928. With the onset of the Great Depression, unions and their influence declined.

World War II marked an even greater decline in workers' rights and working conditions. Several laws were passed to mobilize and regiment the work force in support of the war effort. These included Kokumin Chōyō-rei in 1939 (Personal Service Drafting Law); Chingin Tosei-rei in 1939 (Wage Control Ordinance); Kokumin Kinrō Fukoku Kyōryoku-rei in 1941 (National Labor Patriotic Cooperation Law); and Kokumin Nōryoku Shinkoku-rei in 1943 (National Ability Report Law). Under these laws, people were forced to work in factories with hazardous work environments for low wages and long hours. Unions were disbanded and replaced with a government-organized "national labor front"; regulations setting minimal working conditions were nullified. Workers typically spent twelve to thirteen hours a day, six days a week at their jobs. Rates of labor accidents, tuberculosis, absenteeism, and job desertion all increased. Women and old men were sent into plants to supplement the work force. Employers resorted to harsh discipline to meet production schedules. Labor Administration departments, set up to detect workers' unpatriotic actions and to quell labor disturbances, were part of each local police organization (Tsurumi 1978). Working conditions got even worse toward the end of the war as the military regime became more desperate and Japanese factories became targets of U.S. bombing. Takeshi Fujimoto (1967) estimates that in 1945 alone, hundreds of thousands of workers (including drafted students) were killed or injured as a result of military action against factories.

Also during this period, the Workmen's Pension Insurance program was established. It was said to be designed to provide pension security for retired workers through forced saving. The real purpose, however, was to collect funds for the war effort since no one was to be eligible to receive full benefits for twenty years from the start of the program (Fujita 1984).

Japan's rapid prewar economic growth was dramatically stopped with the devastation marking the end of World War II. The extent of this devastation is difficult to overstate. Edgar McVoy (1955) notes that by the end of the war, 25 to 90 percent of sixty-six major cities had been destroyed. The industrial sector of the economy was especially devastated because of efforts by the American forces to destroy the military capabilities of the Japanese.

The cessation of hostilities put an end to the extremes in working conditions. Philosophies concerning working conditions were significantly liberalized during the U.S. occupation following World War II (Tsurumi 1978). The postwar era, which began at the end of 1945, marked the fourth period of industrialization. One of the major goals of the U.S. occupation forces under General Douglas MacArthur was to get the economy moving again, at least to the point of preventing massive starvation or the need for expensive subsidies from the United States. This was a major task since industrial production in early 1946 was at only 18 percent of prewar levels (McVoy 1986). With U.S assistance and tremendous effort by the Japanese, it did not take long for the Japanese economy to begin to recover. In retrospect, U.S. reconstruction policies proved more successful than anyone could have reasonably expected at the time.

Another major goal of the occupation forces was to inject democracy and personal freedom into the Japanese political system and the work environment in the hope that Japan would become a less belligerent nation as a result. Included in this plan were land reform, women's suffrage, the legalization of independent union activity, the prohibition of racial discrimination, and the dissolution of the *zaibatsu* or cartels, which had holdings and interests in many sectors of the economy (McVoy 1986). It was believed that the *zaibatsu* wielded undue influence in setting national policies, especially regarding Japanese expansionism in pursuit of the "Greater East Asian Co-Prosperity Sphere." As with some other reforms, the actual implementation of the dissolution policy proved to be less extensive than initially planned and the Japanese government used the many remaining *zaibatsu* and large corporations as engines of economic recovery (Pempel 1990).

Another thrust of the occupation policies was the breakup of the labor boss system prevalent in certain industries such as longshoring, forestry, and construction. Labor bosses contracted with employers to supply workers on a temporary basis (in some cases, daily). Individuals interested in work had to go through these bosses, who took a large portion of the employees' wages for themselves. The bosses enforced the system by using threats and strong-arm tactics. The occupation forces saw that the labor boss system was part of the overall coercive, undemocratic Japanese society that they aimed to change. Their efforts to bring about change were not completely effective (McVoy 1985).

New laws, dramatically affecting labor and working conditions, were enacted beginning in 1946, filling a virtual vacuum. The Ministry of Labor was also established during a brief period of ascendancy by the Socialist party shortly after World War II. In the newly formed postwar goverment, this cabinet department of the executive branch had responsibility for overseeing the enforcement of the various laws and regulations (many newly enacted) pertaining to the work environment.

The Labor Standards Act of 1947 addressed workplace issues such as hours of work, safety regulations, industrial accident compensation, and the prohibition of specific categories of workers (women and minors) from night work. The Trade Union Act of 1945 (amended 1949) and the Labor Relations Adjustment Act (1946) legalized trade unions and their activities. The Employment Security Law (1947) outlawed discrimination and guaranteed workers freedom of choice in employment. Taken together, these laws formed the framework for modern Japan's labor and management relations (Tsurumi 1978). In addition, the 1947 Workmen's Compensation and Insurance Law established a national insurance scheme (Patrick 1976). Systems for permanent disability pensions and long-term sickness payments followed during the 1960s.

With the legalization of trade unions after the war, union membership increased dramatically in a brief period. By the end of 1946, union membership had risen to 3.7 million, or 39.5 percent of the work force, compared with a prewar high of 421,000 members in 1936, or 6.9 percent of the work force (Japan Institute of Labour 1983). Accompanying this increase in membership was a rise in union militancy, leading to labor-management conflict over the control of major corporations. This conflict was marked by frequent strikes and violence and in essence was class warfare. Eventually, though, management, with the help of the government and the U.S. occupation forces, broke the back of the militant union movement; organizations and workers who were more willing to cooperate replaced the unions and union leaders.[1]

Demographic, Geographic, and Economic Conditions

Several demographic and geographic conditions have significant implications concerning a broad range of economic and organizational issues in Japan. One such condition is the extremely high population density. Japan has a land area about the size of California and a population approximately half that of the United States. The result is a population density of more than eight hundred persons per square mile, more than thirteen times that of the United States. This figure understates the degree of crowding, however, since approximately two-thirds of Japan's area is mountainous and uninhabited. Furthermore, the land that is available is notoriously lacking in natural resources, requiring the Japanese to import raw materials for most production. This land and resource poverty has contributed to a national sense of insecurity among the Japanese that has affected their interactions with the outside world.

For several reasons, including the high population density and the strong economic growth, the market value of land in Japan is astronomically high. Government support and protection of agriculture ties up valuable land in inefficient, small family farms, some of which are located in major metropolitan areas such as Tokyo and Osaka. Japan's agricultural support policies are undertaken in large measure because of the disproportionate political power of farmers, who have historically supported the dominant Liberal Democratic party (LDP). These policies also forestall farm bankruptcies and the inevitable social unrest that would surely result if the farmers were forced to face world agricultural competition without government support. Further, personal preferences reinforced by government tax policies dictate that relatively little land becomes available for sale; most is passed on through families and corporations. Finally, real estate prices have been inflated by speculative buying.

In the 1980s, interest rates were low and corporate stock prices were very high, so that individuals and corporations had plenty of cash to buy property in anticipation that prices would be even higher in the future. Land prices in Tokyo at the time were approximately one hundred times those in Los Angeles, according to a survey by the Japanese Association of Real Estate Appraisal (*Roanoke Times and World News* 1989). Moreover, real estate experts have estimated that in 1990 the total value of land in Japan (excluding the value of buildings) was $14 trillion, approximately twice that of the United States, which has twenty-five times the area and abundant natural resources (Borrus 1990). In spite of this seemingly rosy picture, the beginning of the 1990s saw a weakening of the Japanese real estate market and fears of a major plunge in prices brought on by a sharp rise in interest rates and a large decline in stock prices.

The price of land obviously affects the construction of new (and safer and more healthful) production facilities. Expansion may require expensive reclamation projects, such as filling in ocean bays or leveling mountains. Many companies thus elect to renovate existing facilities rather than construct new ones. Nonetheless, some companies took advantage of Tokyo's rising land prices in the 1980s by selling off factories in the Tokyo area and building new factories in less densely populated places.

The Japanese have traditionally dealt with the high cost and scarcity of land and raw materials by developing personal habits and business practices that emphasize efficiency and conservation. Their use of small cars, miniaturized electronic equipment, and just-in-time inventory systems, as well as an obsessive concern for quality are among the most obvious ways they have adapted.

Another significant problem that will have increasing importance for Japanese businesses is the aging of the population, which is occurring more rapidly than in any other industrialized country (Brauchli and Chandler 1990). This is the result of a number of factors, including the aging of the postwar baby boomers, a low birth rate, and a longer life expectancy than in any other industrialized country. Moreover, the work forces of major companies are aging at a faster rate than the population as a whole. Many workers were hired in the boom years of the 1950s and 1960s and, given the lifetime employment policies, are still with these companies. To cope with the scarcity of new recruits, businesses are allowing workers who want to do so to continue working beyond the normal retirement age of fifty-five. In addition, increasing numbers of foreign workers, including illegal immigrants, are now filling jobs that no Japanese want to perform, in some cases because of the hazardous nature of the work. This change also threatens the cultural homogeneity that the Japanese prize so highly.

Labor-Management-Government Relations

The years following World War II marked a period of tremendous economic growth and accompanying improvements in the standards of living of Japanese workers. Clearly, external circumstances, including U.S. assistance and purchases to support the Korean War effort, contributed to this growth. Nevertheless, the Japanese economic system itself was largely responsible for this success.

The standard view in Japan is that the government is the planner and financier of economic growth. The Ministry of International Trade and Industry (MITI) supposedly picks likely winners and losers in the economy and targets resources for the winners while shifting resources from the losers. Achievements in industries such as textiles, iron and steel, shipbuilding, and consumer electronics are attributed to MITI's success in targeting. But there have also been notable failures. MITI initially discouraged companies from manufacturing automobiles in the obviously mistaken belief that Japan could never compete with the United States and other developed countries in the production of such a sophisticated product. In addition, it subsequently discouraged companies, including Honda, from entering the market on the grounds that the industry was too small to support additional firms.

Improvements over the years in wages, hours, and conditions of work have resulted from labor-management negotiations as well as regulatory and legislative actions. In general, regulatory and legislative actions have been relatively more important for employees in smaller firms, many of which are subcontractors of larger firms. Employees in larger firms more often have been able to achieve improvements in wages, hours, and working conditions, including those related to safety and health, beyond the minimum government requirements.

In general, government workers and the institution of government are held in much higher regard in Japan than in the United States. Government jobs, particularly in MITI and the Ministry of Finance, are coveted by the top graduates of elite universities. When government workers retire and take a job in the private sector, their entry into a company is referred to as *amakudari*, or the descent from heaven. As we shall see, the higher prestige accorded civil servants has implications for the regulation of workplace safety and health.

The Japanese corporation has long been pictured in the Western press and business publications as a paternalistic organization that cares for workers' every need. Relations between management and the workers or the union that represents them is characterized as cordial and harmonious. Three components of the Japanese employment and industrial relations system, the so-called three sacred treasures or three divine treasures, have been credited both in Japan and abroad with promoting Japan's rapid economic growth. These are the lifetime employment system, the *nenko* wage system (i.e., wages are based on age and length of service), and enterprise unionism (i.e., unions are organized on a company basis, which means they are more cognizant of and concerned about the needs of the company than are unions organized on an industrywide or economywide basis). This description accurately portrays Japanese labor-management relations to a point, but beyond that, it is essentially a stereotype (Shimada 1980). It more closely describes conditions of permanent employees in large master employer firms, but it is not completely accurate even for them. It even less accurately describes temporary workers and workers in subcontractors and medium and small companies.

Master employers are companies such as Toyota, Nissan, Honda, Sony, and Toshiba that are familiar names in the West. Workers in these companies receive good wages, have lifetime employment, work under pleasant conditions, and participate in decision making, all of which have been cited as contributing to the success of Japanese industry. These employers also provide other services not usually offered by Western companies. Company-owned housing and hospitals are common, for example. The hospitals serve the needs of the workers and their families, and some provide services to the wider community on a for-profit basis. The company typically provides financial assistance at major points in an employee's life, such as when he or she begins employment, buys a house, and gets married. Company-sponsored clubs, vacations, exercise programs, and other activities provide for a wide range of social needs. The recent trend among corporations to erect memorials to deceased employees has extended the employer-employee relationship even further (Ueshima 1989).

In return for providing for workers' needs, the company expects a high degree of loyalty from its workers and cooperation from the unions. This relationship mirrors relationships in feudal times, when subjects were expected to show extreme loyalty to their lords in return for the benefits bestowed upon them. Work is expected to be the primary focus of the individual's life. Voluntary overtime in the form of participation in work-team meetings is expected of production workers. White-collar workers normally stay at work well beyond the official end of the day and often spend their evenings socializing with co-workers, all the while discussing business matters. Workers are usually expected to work voluntarily through at least part of their vacations and to be at work except under extreme circumstances such as a death in the immediate family or a serious illness or injury that makes coming to work a physical impossibility. Workers are also expected to stay with the company until retirement.

Lifetime employment is something of a misnomer even for permanent employees in large master employer firms. Except for top-level managers, who may stay on beyond the official retirement age, many retired workers must enter the job market again to meet their financial needs because their pensions are inadequate. Great efforts are made to ensure continuous employment of permanent workers until retirement, but economic downturns and failures of individual firms make it an impossibility for all permanent employees. Moreover, job-hopping, virtually unheard of ten years ago, is now on the increase as workers voluntarily forgo the security of lifetime employment in pursuit of more rapid advancement.

Compensation based on age and seniority is a standard practice in Japan, but pay for performance is often part of the total compensation package. In Jidosha, for example, 55 percent of one's wages are based on age and 45 percent on ability and skill.

Although lifetime employment is considered a major benefit for permanent workers, it also gives companies enormous power over workers. To be involuntarily dismissed is not only devastating financially but a major loss of face. Given the restricted numbers of opportunities for older workers (i.e., anyone older than a recent college graduate) and the stigma associated with being fired, it is highly unlikely that a dismissed worker would be able to find comparable employment elsewhere. Viewed in this way, the compliance, obedience, and dedication of Japanese workers to their companies is not simply a cultural artifact; it has economic origins as well. Some pundits refer to Japanese management techniques as Theory F, that is, management by fear, in contrast to the flattering Theory Z management approach postulated by William G. Ouchi (1981).

Unions in Japan are more sympathetic to management's positions in labor negotiations, and they make an effort to understand the company's needs and financial constraints. The response of the auto company unions to plans to locate production in the United States is typical. Some unions initially resisted the plans (Halberstam 1986), but most eventually supported the idea out of political and financial necessity. According to the president of the Jidosha union, the company would have suffered financially and even more union jobs might have been lost if the union had not cooperated in moving operations to the United States.

One factor that contributes to labor-management cooperation is the ability of union officials to move into management positions. Although they are likely to drop their union membership as they move up the management ladder, union members often fill lower-level supervisory positions. In contrast, a union official moving into management is a relatively rare occurrence in the United States. The ability to move into management helps promote mutual understanding of labor's and management's positions because some individuals are able to view contentious issues from both sides, and, in many cases, labor and management leaders are friends.

Japanese unions clearly have made significant advances since the end of World War II in job protection and wage increases. Unions have also influenced the distribution of wage increases within enterprises, so that the overall distribution of wages is more egalitarian (Shimada 1980). Nevertheless, there is a down side to the close labor-management relations in Japan. Individual workers who disagree with established company and union-endorsed policies are afforded fewer protections. In fact, on the rare occasions when a dissident speaks out against or acts contrary to such a policy, the union is as likely to discipline the worker as the company is. This situation obviously places greater curbs on individual rights and freedoms in the workplace than in the United States.

Much of what has been written about the Japanese economy and about labor-management relations in Japan has focused on permanent workers in the well-known and highly successful master employer companies. Master employers employ only about 30 percent of the work force, however. In other important sectors of the economy, such conditions do not prevail. One such sector is composed of small or medium-sized subcontractors of the master employers. These companies are under extreme pressure from master employers to control costs, and they are also forced in various ways to bear the brunt of fluctuations in demand so that the master employers can maintain lifetime employment policies. Moreover, they are frequently called upon to provide employment for employees of the master employers who become redundant or who need to be reassigned in order to maintain the seniority-based system (so that a more senior worker will not be supervised by a less senior worker). In addition, strict adherence to the Japanese just-in-time inventory system often requires subcontractors to bear much of the cost of maintaining inventories for master employers. Because of the unfavorable conditions imposed on the subcontractors, they are unable to provide the same degree of benefits or the favorable working conditions offered by master employers.

Secondary classes of workers also exist within the master employers. Temporary workers who are hired to meet surges in demand and are then laid off are one such group. Their wages are close to those of permanent workers, but they receive very few of the fringe benefits that comprise a significant portion of the total compensation package of permanent workers.

In most cases, women also comprise a secondary class of workers. They are expected to work only a few years until they get married and then to quit work to raise their children. Increasingly, however, older women are reentering the job market after their children are raised. Even college-educated women are put in noncareer tracks with much lower pay and benefits. The Equal Employment Opportunity Law, which prohibits discrimination on the basis of sex, was passed in 1986, but so far it has resulted in only minor improvements in the status of women at work. The increasing labor shortage may open up more opportunities for women, but significant cultural obstacles will need to be overcome for this to occur.

Although many features of the Japanese employment system are often exaggerated, it clearly has operated to the advantage of Japanese firms in international competition. Both the historically low cost of labor (now a thing of the past) and high productivity have provided significant competitive advantages to Japanese firms. So too has the cooperativeness of labor. This cooperativeness is reflected in the statistics on industrial disputes. As table 2-1 shows, the average number of days lost

Table 2-1. Total Workdays Lost Because of Industrial Disputes, per Thousand Workers, 1975–88, in the United States and Japan

	United States	Japan
1975	204.58	153.47
1976	269.98	61.73
1977	231.02	28.43
1978	247.52	25.10
1979	483.08	16.98
1980	209.90	18.09
1981	168.41	9.92
1982	91.04	9.54
1983	173.16	8.84
1984	80.94	6.14
1985	66.07	4.55
1986	108.22	4.33
1987	39.74	4.33
1988	38.11	2.89
Average	172.27	25.31

Sources: International Labour Office 1983, 1986, 1990.

to industrial disputes over the years considered (1975–88) was 25.31 days per thousand workers. This is approximately one-seventh the number lost per thousand U.S. workers (172.27 days) during the same time.

In general, unions do not play as great a role in Japanese society as they do in the United States despite the somewhat higher overall rate of unionization in Japan (25 percent versus 16 percent). One coordinated activity among Japanese unions is the annual spring wage negotiations, which are referred to as the Shuntō or Spring Offensive by labor. Various union associations coordinate the negotiating strategies of member unions, which engage in separate negotiations with management counterparts. With the decline in union membership in recent years, the Spring Offensive has not been particularly successful. Traditionally, union associations have also provided advice and assistance to member unions regarding occupational safety and health and have offered forums for representatives of the member unions to get together to share ideas regarding safety and health activities.

Union associations, organized along industry lines, coordinate union bargaining activities as well as other activities in the political arena. Union associations in turn have formed national union confederations. As of the mid–1980s, there were four of these confederations: Dōmei, Sōhyō, Chūritsurōren, and Shinsanbetsu. Together they represented approximately 66 percent of Japan's total union membership (Japan

Institute of Labour 1983). More recently, however, there has been a major consolidation of the confederations.

Dōmei, Sōhyō, Chūritsurōren, and Shinsanbetsu were differentiated in part by the industrial sectors in which their members worked but even more by their political orientation. Sōhyō was composed largely (68 percent) of public sector workers in government services such as education, telecommunication, postal service, and transportation. Its primary distinction, however, was its support of the "anti-capitalist/anti-imperialist" policies of the Japan Socialist party. Sōhyō's principal antagonist within the labor movement was Dōmei. It was composed largely of private sector workers (93 percent) and was more moderate in political orientation. It supported the middle-of-the-road Democratic Socialist party. Chūritsurōren, which was composed solely of private sector workers, and Shinsanbetsu, which was the smallest of the four confederations, followed less distinctive political lines.

In 1987, Dōmei, Chūritsurōren, and Shinsanbetsu were dissolved and member unions joined together to form a new confederation called Rengō (Japanese Private Sector Trade Union Confederation). Most of the private sector unions in Sōhyō also joined. In 1989, Sōhyō was dissolved, and its affiliates also joined Rengō (U.S. Department of Labor, Bureau of International Labor Affairs 1989). With such a diverse membership, it is unclear how Rengō will function or what its influence will be.

Employers are also organized into various confederations that represent them politically and that can negotiate broad understandings with union confederations. These confederations are at the industry level. At a higher level, there are five pinnacle organizations: Nisshō (National Chamber of Commerce and Industry of Japan), Keidanren (Japan Federation of Economic Organizations), Keizai Dōyūkai (Japan Committee for Economic Development), Nikkeiren (Japan Federation of Employers' Associations), and Sanken (Industrial Problems Study Group). Nikkeiren, with a membership of thirty-one thousand corporations and fifty-five industry associations, is the most significant of these organizations (U.S. Department of Labor, Bureau of International Labor Affairs 1989).

Although most individual unions do not exert much influence on Japanese politics, Sōhyō and Dōmei were quite active politically, usually electing union members to fill a high percentage of the Diet seats held by their affiliated parties. Many private unions and industry federations of unions lack such political clout since they largely support the political activities of their respective employers and employer organizations. The Federation of Japan Auto Workers' Unions (Jidōsharōren), for example, which is an independent federation, has few if any significant differences with the stated positions of its corresponding management organization, the Japanese Automobile Manufacturers' Association (JAMA). This blurring of labor-management positions was even evident in Dōmei.

Sōhyō and its member unions engaged in activities that were very much at variance with the stereotypic Western view of compliant Japanese unions. The controversy surrounding the privatization of the Japan National Railway, though perhaps an extreme case, serves as an example. Because of the privatization, which took place in 1987, thousands of workers were laid off. Even though the government undertook major efforts to place these workers with other employers, the workers and the union fiercely resisted privatization. Trains were sabotaged, and a union official thought to be too weak in his opposition to privatization was murdered by militant union members (Burgess 1987). Further, according to the news media, nearly one hundred soon-to-be-displaced workers committed suicide in 1986.[2]

Sōhyō was a strong critic of the ruling Liberal Democratic party. Domestically, it fought for expanded labor rights in such areas as wages, working conditions, working hours, and the right to strike; for a nationally uniform minimum wage; and for protection for workers from dismissal. On the international front, it strongly opposed Japanese cooperation with the United States on military matters and advocated worldwide nuclear disarmament. It also protested many U.S. policies in the Pacific region, including, most notably, the Vietnam War.

Political and Governmental Conditions

Japan has a parliamentary form of government with two houses in the legislative branch or Diet. The ruling party forms the executive branch, and ministers of the cabinet are appointed by the prime minister. The prime minister and the cabinet ministers also serve in the Diet. Although the government has a federal structure whereby the country is divided into forty-seven prefectural jurisdictions, each with its own government, most of the power resides in the central government.

Except for six months when the Socialist party ruled, the Liberal Democratic party and its predecessor parties have been in power since World War II. Within the LDP, many branches or coalitions have had their times of ascendancy and their times of decline. Ministerial appointments are often the price of support by one faction to allow another faction to rule. Prime Minister Yasuhiro Nakasone, who served from 1982 to 1987, was the first prime minister to pursue a highly visible leadership role. Previous prime ministers were willing to exert influence behind the scenes or to be consensus builders among the various factions.

Career civil servants wield a great deal of influence in their respective ministries. This is because of the rapid turnover of cabinet ministers, a result of the short tenure of governments, and the ministers' heavy commitments to political activities and Diet proceedings. The authority of civil servants is further enhanced by the high prestige attached to their work. Much of this influence is informal. It develops through bonds of personal friendship with industry leaders, often during college. We shall see how the Ministry of Labor influences the regulation of working conditions.

The emperor is the figurehead leader of Japan, and although Hirohito renounced his divinity following Japan's defeat in World War II, some Japanese still think the emperor is of divine origin. Years are numbered by the year of the emperor's reign, so that 1992 was Heisei 4 or the fourth year of Emperor Akihito's reign (*Heisei* means "achieving peace").[3]

The death of Emperor Hirohito reignited debate about the role of the emperor. Concerns were raised both in Japan and abroad that the Shinto burial rites marking his death violated the constitutional provisions of separation of church and state imposed on Japan by the United States. Fears were also expressed that right-wing nationalists were attempting through the burial rite to revive the myths of imperial divinity and racial supremacy that undergirded Japanese militarism in World War II.

Japan is currently entering a critical era. The rise in the value of the yen has produced a crisis, referred to as the *endaka* (literally, the high-yen condition). It also seems clear that the days of dramatic economic growth are over. Economic growth is predicted to average 3 to 5 percent in the foreseeable future. This is far lower than the double-digit rates of the 1960s and also lower than the more moderate growth of the 1970s, although still higher than that predicted for the United States and Europe. Nevertheless, there is concern in Japan about the "hollowing" of the economy as manufacturing operations are moved overseas in search of cheaper labor. One stunning example of this concern was the announcement by Honda in 1987 that it was planning to manufacture automobiles in the United States for export back to Japan because U.S. production costs were expected to be lower. As Japan entered the 1990s, the significant declines in the stock market and the weakening of real estate prices led to fears in Japan (and hopes among some in the West) that the bubble of the Japanese economic miracle might soon burst.

External pressures imposed on Japan as a result of its stunning economic success are also a concern. In addition to being the largest creditor country and home to seven of the ten largest banks in the world, Japan disburses more nonmilitary foreign aid than any other nation. With its continuing massive trade surpluses and mushrooming overseas investments, Japan's economic policies have repercussions throughout the world. As a result, Japan's trading partners are increasingly putting pressure on Japan to assume greater leadership in promoting international economic development and to modify its economic and trade policies. At the same time, its Asian neighbors fear Japan's growing power.

Japan's external political pressures combined with its domestic economic developments are likely to create tensions among labor, management, and government. Whether the predominantly cooperative relations among these parties can continue under these conditions is an important question the Japanese must face. It is also a question of great interest to Japan's competitors and neighbors.

Motivations and Determinants of Corporate Social Responsibility

Given the major differences in social, political, cultural, economic, and legal influences on corporate social responsibility in Japan and the United States, it is not surprising that the development of CSR has been different in the two countries. Figure 1-1 can be used to describe the situation in Japan, particularly the motivations, emphasis, and implementation of CSR. Perhaps the most obvious manifestation of the difference between Japan and the United States is the nature of the interactions between the corporation and the various stakeholders depicted in figure 1-1. Relative to the United States, the social, ethical, legal, and regulatory environments in Japan are tame and predictable. Using F. E. Emery and E. L. Trist's model of environmental dynamism (1965), the environment in Japan can be characterized as placid compared with the reactive or turbulent environment in the United States.

The social unrest in the United States during the mid–1960s and early 1970s generated considerable interest in CSR and pressure on firms to respond to numerous demands. These demands became organized in various movements that continue in one form or another today, including the civil rights movement, the anti-war movement, the women's movement, the consumers' movement, and the environmental movement. These movements, which have presented formidable challenges to U.S. corporations, have, for the most part, only weak counterparts in Japan. Japanese industry also has much more stable and predictable relations with labor and government regulators than do U.S. firms.

Probably the most significant social movement in Japan has been the environmental movement. It emerged during Japan's high-growth phase out of concern about the environmental effects of rapid economic growth. This concern is certainly understandable given the great concentration of industry and the high population density. Moreover, the Japanese have long professed a great, even religious reverence for nature, although the disregard for the environment shown by many Japanese in public places makes one wonder about the sincerity of such professions.[4]

Japan, like the United States, has had its share of environmental disasters. Among the more notable were an outbreak of Minamata disease, caused by industrial dumping of untreated mercury wastes into fishing waters in the Pacific, and of Itai Itai disease, caused by cadmium pollution. In both cases hundreds of people were affected. There has also been a long-standing environmental cause célèbre: the original construction and continued expansion of the Narita Airport outside Tokyo. The enthusiasm (some would say fanaticism) and persistence of the opposition to the airport notwithstanding, Japan does not have an organized national environmental movement comparable to the one in the United States. In fact, victims of environmental tragedies who have lobbied hard for compensation and recognition of environmental hazards have been treated as social outcasts by their fellow citizens, who resented them for blaming their problems on corporations that provided income and jobs.[5] Japan has made notable improvements in the environment in recent years, however, although they were as much the result of foreign criticism as internal pressure (Beck 1986).

Minority rights movements in Japan represent little threat to corporations, largely because minorities represent a minuscule portion of the population (about 0.5 percent). Moreover, in that discrimination on the basis of race is generally an accepted social practice, discrimination by businesses, although illegal, does not create much adverse reaction among the general public. It is difficult for average citizens to be upset about checks by employers of the racial background of job applicants when so many of them hire detectives to check the racial purity of prospective sons-in-law and daughters-in-law.

Economic relations with South Africa, a major civil rights issue in the United States, is scarcely considered an issue in Japan. (Recently, international criticism of Japan for being South Africa's largest trading partner has caused some politicians to take notice of the issue.) There is, of course, no natural constituency in Japan to protest economic relations with South Africa as there is in the United States. The absence of a black constituency also explains how a number of firms in Japan could feel free to use grotesque caricatures of blacks in their marketing activities (Shapiro 1988).

The women's movement in Japan is equally ineffectual. Despite the passage in 1986 of the Equal Employment Opportunity Law, the status of women in the Japanese labor market is probably inferior to that of American women in the 1950s. Moreover, it is not at all clear that significant changes are in the offing. Most Japanese women seem resigned to being second-class citizens.

Stockholders also pose a weak force for change in Japanese corporations. With the threats and distractions posed by dissatisfied stockholders, ranging from "ethical investors" to corporate raiders, CEOs of U.S. corporations might envy their Japanese counterparts. Japanese corporations often employ thugs to intimidate dissident shareholders at annual meetings. In addition, extensive cross-ownership of stock, whereby related corporations own stock in each other's companies, further minimizes the influence of private shareholders.

The consumer movement in Japan is also relatively weak and was so even in the early days of economic reconstruction when products were notoriously poor. Probably everyone has heard stories of the exorbitant prices the Japanese pay for consumer goods, especially those that are imported. These prices are one piece of evidence that consumers' interests are a relatively low priority in Japan. Protectionist trade policies over the years further demonstrate the predominance of industry interests over consumer interests.

Japan also has a lengthy and complex distribution system involving many tiers of wholesalers and distributors and a great many labor-intensive small shops. It would not be unreasonable to speculate that the government may actually encourage inefficiency in the distribution chain as a way to control unemployment. Workers who are displaced because of the ever-increasing efficiency of export-oriented businesses can be reemployed in this sector of the economy, and it is protected from competition by local ordinances that give shop owners a say in whether new, large, and presumably more efficient stores may open in their jurisdictions.

Another reason for the relatively nonthreatening external environment in Japan is the greater degree of insularity of Japanese businesses. Yonosuke Mafune (1988) notes that Japanese corporations have traditionally provided such "welfare facilities" as hospitals, gymnasiums, housing, roads, and utilities for their employees because local governments historically could not afford them. Thus, there was less need and occasion for businesses to interact with local communities.

In sum, businesses in Japan have traditionally faced a much less threatening environment than do businesses in the United States. Moreover, the pressure for social change has weakened in recent years with the slowdown of economic growth brought on by the energy crisis and the sharp rise in the value of the yen. Stakeholders have shifted their attention back to traditional economic concerns. It is reasonable to expect, though, that increased prosperity will lead to greater demands concerning CSR. The Recruit Cosmos influence-buying scandal, which brought down the government of Prime Minister Noburu Takeshita in 1989 and led to the arrest or resignation of a number of prominent business and government figures, is also likely to result in pressure for change. Since it was revealed that large payoffs (in the form of opportunities to buy shares of companies before they were publicly available) were being exchanged for favorable treatment by officials, the general public has perceived a greater degree of inequality in the system than before and is less satisfied with the status quo.

In contrast to the environment at home, Japanese firms find operating in the United States much more challenging. Japanese firms that have made major blunders include but are not limited to Sumitomo Bank and Honda, which had to pay sex and/or race discrimination settlements to employees; Hitachi, which got caught in an FBI sting operation trying to buy IBM trade secrets and was then criticized for trying to buy public favor when it set up the largest Japanese foundation in the United States; Suzuki Motors, which has had to endure ridicule from virtually every U.S. comic because of the rollover problems with its Samurai utility vehicle; Japanese manufacturers of all-terrain vehicles, which have been involved in numerous product liability suits; and Toshiba, which was slapped with trade sanctions and had one of its radios symbolically sledgehammered by U.S. congressional representatives on the steps of the Capitol Building after its subsidiary sold restricted defense technology to the Soviet Union.

Ethics, Social Norms, and Enlightened Self-Interest

As I noted in chapter 1, CSR may be motivated by enlightened self-interest, ethical considerations, or both. Enlightened self-interest is likely to be the dominant motivation for CSR, although one may expect to see altruistic motives occasionally. In Japan, the overwhelming predominance of enlightened self-interest is striking even to the casual observer. One noteworthy example was the Japanese reaction to former prime minister Nakasone's comments about the low intelligence level of U.S. minorities. There was little discussion of the accuracy of this view or of the ethical implications of the racist attitudes it reflected; rather, the predominant concern was the effect the remark would have on Japanese companies doing business in the United States. As we shall see, enlightened self-interest is also the primary motivation for undertaking occupational safety and health activities.

The predominance of enlightened self-interest over ethical standards is a reflection of the dominant moral standards in Japanese society. There is little distinction in Japan between ethics and social norms. This view is consistent with a philosophy of education that emphasizes rote learning and socialization. Even ethics courses make no attempt to provide students with tools for ethical decision making. Ethical theories are studied for their intellectual and historical value. They need not be applied because Japanese society or one's family, school, or business tells one what is right and what is wrong. Following rules is right; breaking rules is wrong. Individualism or, more pejoratively, egoism (*kojinshugi*), which would entail, among other things, deciding for oneself what is right and what is wrong, is strongly discouraged.

Even questioning the rules is not considered appropriate. A lifelong Canadian resident of Japan who was planning to protest the country's onerous requirement that all foreign residents be fingerprinted every few years was warned against doing so by the owner of the university where he taught. (Many private Japanese universities are owned by individuals.) Although the owner could understand his grievance and the silliness of the rule—one's fingerprints are not likely to change—he was concerned that students might get the idea that it was appropriate to question rules, regulations, and laws.

The Japanese view of ethical reasoning is consistent with several dominant aspects of Japanese culture. These include the sense of group membership, the sense of duty based on personal relationships, and the sense of loyalty. These characteristics of Japanese culture have shaped the social pressures noted earlier and shape the way CSR is implemented.

Personal identity in Japan has traditionally been defined on the basis of group affiliation, most notably one's family and place of employment. It is only logical that social norms would play a dominant role in moral reasoning.

The sense of group identification also produces a we-versus-them attitude that in its broadest manifestations results in the sense Japanese have of being unique among races and of being in economic and cultural conflict with the rest of the world. Long-standing and successful efforts to restrict immigration to Japan is one consequence of this outlook. So too is the exclusion of minorities from the mainstream of social intercourse. On a more micro level it results in the sharp distinctions between those who are considered full-fledged members of a corporation and those who are not.

The group orientation also explains the reactions Japanese males have to women and minorities in the United States. (Some would say that even white males are treated as outsiders by the Japanese.) Working in the United States requires a whole new way of acting, especially toward women and minorities, to avoid running afoul of the law and social norms. No doubt, this is very difficult for some Japanese managers who are working in U.S. operations.

Confucian ethics define morality in terms of an individual's duties in specific relationships (father-child, husband-wife, oldest brother–sibling, friend-friend, and, most importantly, master-servant). The relationship, not an abstract ethical principle, is the key to prescribing behavior. Thus, there is no generalized sense of charity; one has a duty to help people, but only those with whom one has a relationship. The Western notion of charity derives from Judeo-Christian traditions, which provide a rationale for individual and collective altruism.[6] Loyalty to one's lord or master is a very strong tradition in Japan, dating back to feudal days. Samurai followed a code called Bushido, vestiges of which, like chivalry in the West, have survived (Whiting 1977). In addition to loyalty, Bushido emphasizes values of self-discipline, unquestioning obedience, self-denial, and stoicism, all of which have implications for safety and health management. Loyalty to one's lord was supposed to be unconditional, overriding other loyalties and duties. In a popular play (*Tōjin Okichi*), the hero is forced to sell his daughter into prostitution to fulfill his responsibilities. His sacrifice is viewed as a highly honorable act.

Loyalty is still highly valued today. The feudal lord has been replaced by the feudal corporation which provides all the necessities of life and guidelines for acceptable behavior. Loyalty to the emperor was considered absolute and not conditioned on merit or righteousness; so too some Japanese now consider loyalty to their employer to be absolute. It take precedence over other duties, even to one's family and (in extreme cases) to following the law. This orientation explains several national and corporate scandals that caused considerable consternation in Japan and the United States.

One such scandal was the Lockheed payoff, which brought down the administration of Prime Minister Kakuei Tanaka. In this case, several employees of All Nippon Airways were found guilty of breaking Japanese laws by taking bribes from Lockheed officials. Although they suffered some criminal penalties, the employees were welcomed back warmly to the corporation. Their loyalty to the corporation was more important than their adherence to an abstract moral or legal principle, and their loyalty effectively exonerated them among their co-workers.

In the more recent and more serious Toshiba case, employees sold restricted defense technology to the Soviet Union to benefit their organization, even though the act was highly inconsistent with national interests and international obligations. Several corporate officials resigned as a result of the incident, but many in the United States regarded the resignations as mere tokenism. Consequently, the U.S. Congress imposed trade sanctions on Toshiba.

The employees who are responsible for such fiascoes are viewed as having been misguided in their sense of propriety. Nevertheless, they are considered honorable in that they attempted to further their companies' interests at considerable personal risk. Although some in the United States might have a similar view of corporate criminals, in general we place less value on loyalty when it comes at the expense of obeying the law or acting in accordance with ethical standards.

The within-group/out-of-group distinction also has important implications for how CSR is put into practice in Japan. The distinction affects who is considered a stakeholder and the priority one group of stakeholders has vis-à-vis others. The interests of permanent employees are the primary concern of Japanese corporations. Concerns for other stakeholders and for the wider social environment are assumed to be met simply by following the law or social customs. There is an underlying belief that legitimate social demands will be codified into law or will be accepted by consensus. Corporations need not be concerned with demands of interest groups who are unable to get their viewpoint so codified or so accepted. Thus, minority views receive far less attention in Japan than in the United States. This explains in part the difficulties Suzuki and the all-terrain vehicle manufacturers encountered with consumer groups. For the most part, product safety issues in Japan are worked out between the industry and responsible government agencies; autonomous consumer groups have relatively little input. Consumers who are injured as a result of a product defect usually are satisfied with an apology and what we would consider token compensation.

The Japanese penchant for conformity and compliance ensures that most laws are followed, at least in a *tatemae* sense, even if there are no punitive sanctions for noncompliance. Thus, many Japanese laws and regulations carry no penalties for violation. The recently passed sex discrimination law is one example. The law is intended to bring women into the mainstream of corporate life, but for a multitude of reasons, relatively few women have benefited, despite the widespread contention that discrimination has been effectively abolished.

As we can see, the way CSR is put into action in Japan is relatively simple compared to the situation in the United States. The stable social environment enables firms to promote the interests of stakeholders (with a tilt toward permanent employees) largely through compliance with the law and social custom. In many cases, masterful application of the principles of enlightened self-interest further promotes stakeholders' welfare and corporate profits. In-depth consideration of the management of a specific CSR issue, occupational safety and health in Jidosha's Japanese and American operations, will provide additional insight into Japan's management of CSR. First, however, I will consider the regulatory framework in which the management of safety and health in Japan takes place.

3

The Safety and Health System in Japan

This chapter describes the Japanese occupational safety and health system, including legislation and institutional structures that promote safety and health. In addition, comparisons are made with the U.S. system. For background information on the U.S. system, the reader might want to refer to one or more of the following: (Ashford 1976; Lofgren 1989; McCaffrey 1982; Mintz 1984; Noble 1986; Rosner and Markowitz 1987; Viscusi 1983; and Wokutch 1990a.) These comparisons are based on data on enforcement activities at the national level as well as on data on staffing levels. Particular attention is paid to the variability over time of these activities and staffing levels for what they tell us about the stability of national safety and health policies.

Regulation of Occupational Safety and Health in Japan

Early Legislation

Industrial safety and health problems developed in Japan following the Meiji Restoration as Japan rushed to catch up with industrialized countries, but, except for favored groups such as public employees, workers had little or no protection. In a critical account of safety and health conditions during these years, Fujimoto writes:

> In those days, working conditions in Japanese industries were also terribly bad. Working hours in mines were usually 10 or 11 and factory workers worked as long as 11 hours in metal manufacturing industries and 11 to 15 hours in txtile [sic] industries including rest and meal intervals. Holidays of manual workers in all industries were at most two days per month. . . .
>
> Many employers dealt with workers just as slaves, some employers such as mine owners and contractors or sub-contractors in construction outrageously oppressed workers' discontent by violence. The majority of em-

> ployers were entirely indifferent to the safety of workers at the work places. As a matter of course, bad labour conditions as such, deprivation of human rights from workers and employers' indifference to safety resulted in high rates of accidents in those days (1967:3).

The Mine Law of 1905 was the first safety and health initiative that affected the private sector. It was directed at the mining industry, which experienced an especially high rate of fatal injuries at the time. Among other provisions, it established a program for companies to assist injured miners financially.

The Factory Law of 1911 was the first general safety and health legislation. It was directed at private sector factories employing fifteen or more employees (later reduced to ten) in hazardous work (Japan International Cooperation Agency 1985). The Factory Law included specifications regarding worker safety, such as restrictions on employment in harmful and hazardous jobs for workers, notably females and youths under fourteen years of age in protected categories, and limitations on working hours and night work. It also provided for a factory inspection system and established the first general program in Japan to provide assistance to injured workers. Workers in government factories were covered under a workers' compensation system as early as 1875 (Bennett and Levine 1976; Hazama 1976), but the Factory Law provided the first comprehensive coverage. Neither the Mine Law nor the Factory Law referred to the need to provide compensation or reparations to injured workers but rather called for assistance. Moreover, they stipulated that negligence by workers was a legitimate reason to withhold payments (Japan International Cooperation Agency 1985).

The Factory Law is not universally recognized as a significant advance in worker safety and health. Fujimoto, quoting an unidentified foreign writer, noted that it simply "approved officially the cruel exploitation of workers by Japanese employers" (1967:4). Nevertheless, it set the stage for subsequent progress.

The Health Insurance Law of 1922, which became operational in 1927, extended workers' injury and illness compensation benefits further by providing benefits for short-term injuries from both work-related and nonwork-related causes. This insurance applied only to workers covered under the Factory Law, however. Further, it relieved employers of responsibility for providing benefits directly to incapacitated workers to the extent that they were compensated under this plan.

Workers in previously uncovered industries were provided short-term injury and illness compensation by the Workmen's Accident Relief Law and the Workmen's Accident Relief-Liability Insurance Law, both of 1931. These provided coverage for workers in construction and building, transportation, and cargo-handling industries (Japan International Cooperation Agency 1985).

The Workmen's Pension Insurance Program of 1941 provided pensions for workers with long-term disabilities. Certain restrictions on eligibility for coverage under this program were subsequently liberalized and benefits were increased by amendments approved in 1944.

In addition to progress on compensation for injuries and illnesses, the 1920s saw attempts to improve working conditions and provide various amenities to workers in an effort to discourage unionization. International pressure, much of it channeled through the International Labour Organization, also pushed the Japanese toward following the work standards and practices of the West (Patrick 1976). Worker safety and health was given special emphasis in 1928, when the government instituted the first national safety and health week. This tradition continued, although on a lower-key level, even during the Depression, when the influence of unions waned, and during World War II, when independent union activity was forbidden. Despite these improvements, working conditions in Japan remained poorer than in more advanced industrialized countries. Moreover, conditions deteriorated significantly in the years leading up to and during World War II and were terrible by the end of the war.

Postwar Safety and Health Legislation

Following the war, the U.S. occupation forces were appalled at Japan's working conditions and the minimal legislative and regulatory protections for workers. Moreover, the U.S. command generally held the belief that the lack of labor rights had contributed to Japan's militaristic foreign policy, which was supported by influential industrialists. Thus, the U.S. forces rewrote major elements of labor law in the process of restructuring Japanese society. The legislation covered all aspects of working conditions and had important effects politically, as well as a significant impact on occupational safety and health.

The Japanese constitution, drafted by occupation staff and adopted in 1947 by the Japanese, guarantees workers certain rights. Article 27 in a "bill of rights" appended to the constitution states that "all people shall have the right and the obligation to work . . . [that] standards for wages, hours, rest and other working conditions shall be fixed by law [and] . . . children shall not be exploited." And Article 28 states that "the rights of workers to organize and to bargain and act collectively is guaranteed" (quoted in Levine 1981:53). There are no such constitutional guarantees for labor unions in the United States.

The Labor Standards Law, passed in 1947, and subsequent legislation provided specific guidelines for carrying out the broad aims of the constitution. The law is a very comprehensive piece of legislation that includes chapters dealing with general provisions; labor contracts; wages; working hours, breaks, and holidays; safety and health; women and minor workers; training of skilled laborers; accident compensation; rules of employment; workers' dormitories; inspection provisions; miscellaneous provisions; and penal provisions. The chapter on safety and health provides a general framework for safety and health regulation that was not superseded until the Industrial Safety and Health Law of 1972. (The full text of these and other Japanese labor laws are in Ministry of Labor 1980.) Other chapters established a comprehensive industrial injury and illness compensation system, set limits on hours of work, and prohibited specific categories of workers (women and youth) from night work. Nonetheless, certain provisions of the Labor Standards Law, particularly the forty-eight-hour workweek, were widely ignored, especially by smaller firms, at least in the early postwar years (Bennett and Levine 1976).

The Labor Standards Law stipulated for the first time that payments to incapacitated or ill workers were designated as compensation rather than assistance, indicating that it was the responsibility of the employer to provide payments and the right of workers to receive them. The law also provided uniform benefits, rather than payments based on the perceived need of the individual worker. In conjunction with the Labor Standards Law, the Workmen's Accident Compensation Insurance Law established an insurance fund to which employers made mandatory contributions and from which employee injury and illness compensation was paid. Excluded from coverage were the self-employed (unless they elected to participate voluntarily), most public employees, mariners, employees of the national railways (since privatized), and certain employees in agriculture, forestry, and fishery (Japan International Cooperation Agency 1985).

From 1960 onward, improvements were made in the compensation coverage for workers with long-term disabilities and survivors of workers who died of work-related causes. Principal among these improvements was the provision of pensions to replace lump-sum compensation payments, which recipients frequently outlived.

In addition to the democratization policies instituted during the U.S. occupation, technological progress acted as a catalyst for change in Japan's industrial safety and health activities. Innovative uses of raw materials and new processing methods had led to increases in industrial accidents and occupational diseases (Japan Institute of Labour 1982). To cope with this problem, the government launched the first Five-Year

Industrial Injury Prevention Program in 1958. Subsequent programs were begun in 1963, 1968, 1973, 1978, 1983, and 1988. The private sector also promoted safety and health efforts through the establishment of numerous associations at both national and industrial levels.

Organization of Government Safety and Health Activities

Occupational safety and health regulation is under the jurisdiction of the Ministry of Labor. As with other cabinet positions, the minister of labor is an elected member of the Diet and appointed minister by the prime minister. Virtually all officials beneath the minister are career civil servants rather than political appointees, lending a degree of stability to the ministry. There are five bureaus in the Ministry of Labor: the Labor Relations Bureau, the Human Resources Development Bureau, the Women's Bureau, the Employment Security Bureau, and the Labor Standards Bureau. The Labor Standards Bureau is responsible for occupational safety and health matters, along with a range of other workplace issues.

The overall organization of the Labor Standards Bureau is shown in figure 3-1. As it indicates, there are five principal divisions plus two departments: the General Affairs Division, the Inspection Division, the Labor Accident Management Division, the Compensation Division, and the Executive Office for Workmen's Accident Compensation Insurance, as well as the Industrial Safety and Health Department and the Wages and Welfare Department.

The Industrial Safety and Health Department, as the name implies, has primary responsibility for industrial safety and health matters, including participating in the promulgation of safety and health regulations and inspecting workplaces for compliance. The department also advises firms regarding safety and health practices. Other departments and divisions also have safety and health responsibilities, such as the operation of the workers' compensation program (Executive Office for Workmen's Accident Compensation Insurance) and the regulation of hours of work and youth work (Wages and Welfare Department).

Regional (prefectural) bureaus and inspection offices of the Labor Standards Bureau carry out the field inspections for the agency. There are 47 prefectural labor standards bureaus and 348 inspection offices throughout Japan. Together they employ approximately three thousand inspectors who are responsible for overseeing general conditions of work and seven hundred expert officers in industrial safety and health (four hundred safety experts and three hundred health experts).

Figure 3-1. Organization of the Labor Standards Bureau

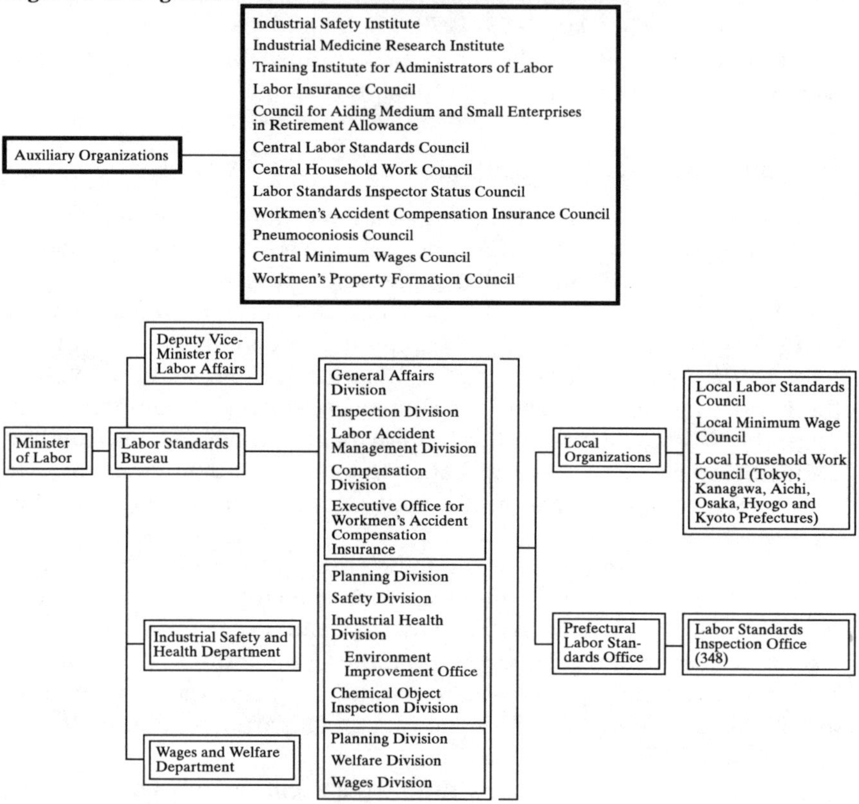

Source: Japan Institute of Labour 1982:31.

Recent Safety and Health Legislation

There are three levels of laws and regulations pertaining to occupational safety and health in Japan. The first level consists of laws that are passed by the Diet. The Industrial Safety and Health Law of 1972 is the most significant example of such a law. Laws passed at this level are the broadest and most general of the three. The second level are cabinet orders. These are issued by the minister of labor with the concurrence of the rest of the cabinet. They dictate, in specific terms, how the laws are to be carried out. The third and most specific level are ministerial ordinances. These provide precise guidelines regarding how certain work activities are to be performed and how various occupational hazards are to be handled. Examples of recent ordinances include the

Ordinance on Prevention of Organic Solvent Poisoning, the Ordinance on Prevention of Ionizing Radiation Hazards, the Ordinance on Safety and Health of Work under High Pressure, and the Ordinance on Prevention of Hazards due to Dust.

Several specific safety and health concerns have been addressed in recent years. The Pneumoconiosis Law of 1960, for example, set up mechanisms to deal with this disease of the lungs caused by inhaling hazardous substances. It is an especially serious problem in the mining industry, where it is better known as black lung disease. The Industrial Injury Prevention Organization Law of 1964 established several voluntary industry organizations to promote occupational safety and health, including the Japan Industrial Safety and Health Association (JISHA). The Industrial Homework Law of 1970 attempted to promote safety and health in the previously largely unregulated and frequently hazardous at-home work environment. The Working Environment Measurement Law of 1975 dealt specifically with procedures for monitoring the work environment for the presence of various hazardous substances. Qualifications of measurement personnel are set forth, and very precise guidelines are given for measuring these substances, including the instruments to be used, the placement of instruments in the workplace, the frequency and duration of measurement, and the length of time for which records are to be kept.

The Industrial Safety and Health Law and the Occupational Safety and Health Act

Japan's most important safety and health legislation enacted in recent years is the Industrial Safety and Health Law of 1972, which, rather than addressing a specific issue as most previous laws had, provides an overarching system for promoting occupational safety and health.

The Industrial Safety and Health Law was passed for several specific reasons. In the late 1960s, a committee composed primarily of academicians and safety and health specialists was convened by the Ministry of Labor to consider the adequacy of the Labor Standards Law. They found that industrial developments and ambiguities in the law had led to an increase in serious accidents and that new legislation was required. (In fact, there had been an increase in serious injuries and fatalities in the 1950s, but they had tapered off in the 1960s; 1961 had been the peak year for both deaths and serious injuries.) Specifically, they noted several problems with the law: (1) there were inadequate guidelines for preventing industrial accidents, which had been exacerbated by the rapid development of industry; (2) mechanisms were needed for providing advice and counsel on safety and health beyond minimum standards of performance; (3) there were too few safety and health specialists; and (4) there were inadequate provisions for coping with the increase in accidents in small and medium-sized companies and in subcontractors (Yoshimoto 1981). Others have noted the need to clarify safety and health responsibilities in cases when employees who worked for different employers worked on the same project (Iino and Seiji 1986). This situation was particularly problematic in the construction industry, but it also existed in other sectors of the economy such as the automotive industry, where subcontractors often deliver supplies directly to work stations on the factory floor.

Testimony in the Diet during consideration of this legislation indicated that a further purpose of the bill was to clarify the line of responsibility for worker safety and health. It was noted that in the past lower-level managers had sometimes been held responsible for accidents for which higher-level managers escaped blame. With this legislation, top management (or the owner if a business was owned by an individual) was assigned ultimate responsibility for workplace safety and health. Responsibility to follow the Industrial Safety and Health Law extended, however, to first-line supervisors and even to individual workers (Yoshimoto 1981).

Officials of the Labor Standards Bureau observed in retrospect that the Labor Standards Law was simply too broad-based in its coverage to deal adequately with safety and health. Only one chapter dealt with this important issue. They also felt it was necessary to streamline and coordinate the various laws and regulations pertaining to safety and health that had been enacted since World War II (Iino and Seiji 1986).

Other industrialized countries had passed similar legislation, and no doubt this had an influence on Japanese legislators. Occupational safety and health protection is a very sensitive issue for the Japanese that affects their national prestige and international trade relations. Clearly, they would not have wanted to be seen as lagging behind other nations in providing for the safety and health of their workers. This would be embarrassing, and it could be used against the Japanese in trade negotiations.

Opinions differ on whether the U.S. Occupational Safety and Health Act passed in 1970 served as a model for the Japanese Industrial Safety and Health Law. There are certain similarities, and the timing seems more than coincidental, even though the Japanese began considering the adequacy of their safety and health legislation before the actual passage of the Occupational Safety and Health Act in the United States. Some Japanese officials believe that British legislation served as the model for the Japanese law, particularly with respect to responsibilities at multi-employer sites. The Japanese form of government has more in common with the British parliamentary structure than with the U.S. federal structure. Moreover, the British tripartite system of labor, management, and government cooperation in the regulation of safety and health (see Wokutch 1990a) is better suited to the overall thrust of labor-management-government relations in Japan than are the confrontational relations in the United States.

The overall purpose of the Japanese Industrial Safety and Health Law is stated as follows:

> To secure, in conjunction with the Labour Standards Law (Law No. 49 of 1947), the safety and health of workers in workplaces as well as to facilitate the establishment of comfortable working environment, by promoting the comprehensive and systematic countermeasures concerning the prevention of industrial accidents, such as taking measures for the establishment of standards for prevention of accident [sic] and health impairment, the clarification of responsibility and the promotion of voluntary activities with a view to preventing industrial accidents (p. 1).

A quick reading demonstrates that this purpose is similar to the one stated in the U.S. Occupational Safety and Health Act (OSH Act) of 1970:

> To assure safe and healthful working conditions for working men and women; by authorizing enforcement of the standards developed under the Act; by assisting and encouraging the States in their efforts to assure safe and healthful working conditions; by providing for research, information, education, and training in the field of occupational safety and health; and for other purposes (Public Law 91–596).

But there are also several differences in the statements. The Japanese Industrial Safety and Health Law specifies that the goal is not just safe and healthful work environments but comfortable ones as well. There are, however, no readily discernible differences in the comfort levels of work environments in the United States and Japan. The encouragement of voluntary activities is specifically cited in the Japanese law but missing from the OSH Act. Voluntarism with respect to safety and health is, in fact, much more greatly emphasized in Japan than in the United States. Another difference, the explicit reference to clarification of safety and health responsibilities in the Industrial Safety and Health Law, is a reflection of the confusion over this issue that the law was meant to rectify.

The purpose statement of the OSH Act also contains several elements not mentioned in the purpose statement of the Industrial Safety and

Health Law. These include providing for research, information, education, and training on occupational safety and health. These items are not specifically mentioned in the purpose statement of the Industrial Safety and Health Law, although they are specified elsewhere as being functions of the Labor Standards Bureau. More significantly, the emphasis in the OSH Act on assisting states in their safety and health activities does not have a direct counterpart in the Japanese law since prefectural governments in Japan have far less autonomy than state governments do in the United States. Prefectural labor standards bureaus are directly accountable to the national Labor Standards Bureau. There is nothing comparable to state-plan OSHAs in Japan.

The Industrial Safety and Health Law includes safety and health provisions pertaining to the following areas: responsibilities of the employers and employees and of manufacturers, designers, and importers of machinery, raw materials, and so on; the development and promulgation of safety and health programs by the Labor Ministry; organizational structures (within companies and plants) for safety and health management; measures to prevent accidents and illnesses to be taken by the employer and the principal employer (where a contractor-subcontractor relationship exists); regulations concerning the use of machines and harmful substances; regulations pertaining to the training and placement of workers; measures to promote the health of workers (including medical exams, special consideration in the placement of middle-age and older workers, and measurements to be taken in the work environment); licensure of workers in certain skilled occupations; the establishment of a comprehensive safety and health improvement program when deemed necessary by the Labor Ministry; periodic inspections of work sites and advance review of plans of new work facilities; and penalties for noncompliance (Japan Institute of Labour 1982).

The stipulation of penalties for safety and health violations in the law is significant since Japanese regulations do not always carry penalties for noncompliance. Maximum fines and in some cases maximum prison sentences are specified for violations of the various provisions of the law, but, as we shall see, prosecution for safety and health violations is relatively rare in Japan compared with the United States. As with other regulatory functions in Japan, the government follows a policy of providing "administrative guidance" rather than serving as a dictator or police officer.

In addition to the responsibilities of the employer, the law specifies responsibilities of the government, company safety and health officials, safety and health committees, the union (where one exists), and individual employees. Two government duties are of particular interest.

These are, first, to provide financial and technical support to small and medium-sized firms so that they can promote occupational safety and health and, second, to train safety and health specialists.

The specific provisions and the overarching philosophy of the Industrial Safety and Health Law differ from those of the OSH Act in a number of ways. I will review the most relevant of these differences.

First, the Industrial Safety and Health Law emphasizes the specifications for industrial structures, whereas the OSH Act emphasizes the legal mechanisms to be followed to ensure compliance with specific regulations, which in turn are expected to promote safety and health.

Second, the Industrial Safety and Health Law and subsequent regulations put more emphasis on a behavioral approach to the promotion of safety and health, that is, modifying the behavior of workers and managers, compared with the OSH Act. The U.S. law emphasizes the use of an engineering-oriented approach, that is, requiring physical changes in the work environment.

Third, the Industrial Safety and Health Law specifically mandates procedures and structures to promote labor-management cooperation on occupational safety and health. The OSH Act notes that cooperation is desirable but stops far short of mandating cooperative arrangements.

Fourth, unlike the OSH Act, which specifies procedures for the review and appeal of standards, the Industrial Safety and Health Law is silent on this issue. There are provisions in both countries for labor and management input into the process of writing standards. In Japan, it is assumed that this process will ensure broad acceptance of the standards; in the United States, such an assumption would be foolhardy.

Fifth, the procedures specified by the Industrial Safety and Health Law (and supplementary regulations) for issuing and appealing penalties are more involved than those specified by the OSH Act. As a result, far fewer penalties are issued in Japan. Penalties are typically issued at the end of extended legal proceedings, however, so that, compared with the United States, the chance of reducing penalties is slim. The method of appeal is also different in Japan, where it is handled by the regular judicial system. In the United States, there is a special Occupational Safety and Health Review Commission.

Sixth, unlike the OSH Act, there are no specific provisions in the Industrial Safety and Health Law for employer or employee representatives to accompany Labor Standards Bureau inspectors on workplace inspections. Nevertheless, employer representatives routinely accompany inspectors, and it is not unusual for an employee representative on the company safety and health committee to go along as well.

Seventh, except for expert officers in industrial safety and health,

Labor Standards Bureau inspectors have the power of judicial police officers. OSHA inspectors have much more limited authority. Japanese inspectors may issue immediate work shutdown orders in the case of imminent danger, whereas OSHA inspectors must secure court orders for shutdowns.

Taken together, these differences indicate that the Industrial Safety and Health Law was designed to promote safety and health through the voluntary cooperation of labor and management with each other and with the government and, compared with the United States, to rely only minimally on legal mechanisms and the use of punitive sanctions while maintaining the threat of swift punishment for egregious violations. These themes will reemerge in the remainder of the chapter as I discuss the safety and health system in Japan in more detail and compare aspects of it with corresponding elements in the United States.

Rule Making, Enforcement, and Assistance Activities

Inspection officers of the Labor Standards Bureau have the right to enter workplaces and question managers and workers about working conditions. They have the right to inspect records, documents, and other articles, as well as the workplace itself, including machines, tools, and raw materials. With their power as judicial police officers, Labor Standards Bureau Inspectors can arrest individuals on the spot for violations of occupational safety and health requirements. Such occurrences are extremely rare, however.

Since the case of *Marshall* v. *Barlow, Inc.* (436 U.S. 307 [1978]), a U.S. employer has had the right to refuse an inspector entry to a work site if he or she does not have a search warrant. Japanese employers do not have this right. The difference reflects the lower priority placed on individual rights and freedoms and the higher priority placed on conformity in Japan as compared with the United States.

Employers in Japan are typically notified in advance of inspections, whereas advance notification is specifically prohibited by the OSH Act. The practice in Japan is consistent with the observed pattern of regulators providing administrative guidance as opposed to threatening punitive sanctions for poor performance.

Consistent with this pattern, fines in Japan are imposed for only the most serious violations. These consist primarily of cases in which serious violations have been uncovered in an inspection and the employer has not corrected them by the time of a follow-up inspection and those in which an accident has occurred because of a serious violation. After an inspector has found evidence that one of these two conditions exist, the case is investigated and, depending on the findings, a local Labor Standards Bureau office may elect to "send the papers" to the prosecutor's office. The prosecutor's office then has the prerogative to pursue the case to trial. Should the office elect to proceed with the case, a conviction is virtually a certainty.

Descriptions of Japanese decision-making processes within organizations apply to decision making at the national level as well, at least with regard to occupational safety and health. New laws or regulations are developed through a tripartite system involving labor, management, and government. (In reality, it could be considered a quadpartite system since academic researchers are also involved.) The government, represented by the Labor Standards Bureau, usually takes the initiative in developing the legislation or regulation. One exception is the legislation on dust in the workplace, which was initiated by labor. The Labor Standards Bureau either conducts its own research on the subject of the legislation or contracts with academic researchers to study the issue and make proposals. Next, a labor standards council with tripartite or quadpartite representation is formed to review a draft of the proposed legislation or regulation. The government plays the lead role in this council. Employer associations such as the Japanese Automobile Manufacturers' Association may be asked to send a representative to the council, and union organizations may be asked to do likewise. Members of the council then attempt to reconcile any differences they have regarding the proposal. In most cases, the interests of these parties are considered in the initial formulation of the proposal and a compromise is reached. Because cooperation is emphasized throughout the process, bickering and criticism regarding safety and health laws and regulations, so common in the United States, are kept to a minimum during consideration and after passage of an act.

One mechanism the Japanese use to ensure that compromises are reached is to prevent potentially contentious issues from reaching formal deliberation until agreement is likely. This approach appears to have been used when labor demanded a reduction in the hours of work. After years of discussion, a consensus has emerged that a shorter workweek would both benefit Japanese workers and improve international trade relations. Reduction of the workweek is now a national goal.

Exposure levels for various harmful substances in the work environment are recommended by a voluntary organization, the Industrial Health/Hygiene Academic Association. Because the levels are recommendations only, the government does not enforce them and there are no penalties for noncompliance. By authority of the Work Environment Measurement Law of 1975, however, the Labor Standards Bureau sets standards for ventilation, stipulates precise methods for measuring substances in the work environment, and licenses experts to perform such measurements.

Decisions regarding what establishments and issues to focus on in inspections are set by area offices and prefectural bureaus of the Labor Standards Bureau in conjunction with the national office. There are regional differences in priorities because of regional differences in the industrial composition of the work force. In one industrial prefecture, companies were said to be targeted on the basis of their injury rates and whether or not they handled toxic chemicals and carcinogens. Although no explicit policy exists, there appears to be a tendency to inspect larger establishments.

Workers may also request inspections. Requests are likely to come from workers in larger companies since workers in smaller establishments tend to refrain from complaining for fear of identification and retaliation. In one recent year, 806 inspections were requested in the industrial prefecture mentioned above. Of these, 477 dealt with pay disputes, working hours, or underage workers. Only 35 pertained to safety and health issues.

In addition to its regulatory activities, the government promotes safety and health activities in other ways. In particular, the Ministry of Labor is involved in training, financing, consulting, and research activities to promote safety and health. A government-funded University of Occupational and Environmental Health was established in 1978, and there are government-sponsored occupational safety and health education centers. The government provides funds to small and medium-sized businesses for safety and health activities through low-interest loans and direct subsidies, and Labor Standards Bureau inspectors provide technical advice to employers to help them improve safety and health conditions and to ensure compliance with regulations. Research activities are carried out in the Research Institute of Industrial Safety and in the National Institute of Industrial Health, which together constitute the counterpart of the U.S. National Institute for Occupational Safety and Health (NIOSH). Other research activities are conducted by the Japan Bioassay Laboratory and the Occupational Health Service Center, which are operated by the Japan Industrial Safety and Health Association.

Workers' Compensation

The workers' compensation system (usually translated "workmen's compensation") is an important element of the occupational safety and health system in Japan. The system is administered in the Labor Standards Bureau and is self-supported by mandatory employer contributions and voluntary contributions by employers and employees not covered by the system but who elect to be covered. Coverage is mandatory in establishments with five or more employees.

As in the United States, employer contribution rates are calculated by type of industry based on the degree of risk. In 1988, rates ranged from a low of .5 percent of wages for firms in the commerce industry to 14.5 percent of wages for firms involved in the construction of new hydroelectric power stations. As in the United States, a firm's rates may be modified depending on its injury and illness experience. Only firms with thirty or more employees are affected by this merit rating system, however.

One difference between the workers' compensation system in Japan and the United States is that, in addition to collecting workers' compensation, injured or ill employees in Japan may sue their employer for negligence. In the United States, this is not permitted, although, as in Japan, third parties, such as suppliers of defective machinery and materials, may be sued. The incidence of such litigation is low in Japan, and the cash awards in successful suits are modest—less than typical traffic accident settlements in Japan and far less than most product liability settlements in the United States (Matsuda 1986).

Compensable injuries and illnesses are specified in guidelines for the workers' compensation system. By and large, local Labor Standards Bureau offices adhere to these guidelines with relatively little controversy. Occasionally, however, decisions at the office level are contested. Workers sometimes question their designated degree of disability, which affects the amount of compensation for which the employee is eligible. The most frequent disagreement pertains to whether an injured individual is an employee and therefore automatically eligible for compensation or whether he or she is self-employed and eligible only for the optional program. Eligibility is particularly difficult to determine in small establishments in which a number of family members work.

Disagreements also arise over claims for injuries and illnesses that are not on the list of designated work-related conditions. The individual may appeal a decision to the prefectural bureau and, if he or she is still unsatisfied, to the national level. The Labor Standards Bureau will appoint a committee of outside experts (the Labor Insurance Appeals Committee) to decide the case. As one might expect, relatively few decisions are appealed. In one typical year more than 257,000 workers' compensation claims were settled, of which only 1,474 (less than 1 percent) were appealed to the prefectural level. Of these, 329 went to the Labor Insurance Appeals Committee. Precise figures were not available, but reportedly very few workers took the next available step and appealed to a committee through the courts (Iino and Seiji 1986).

Because of its far-reaching implications, one appeal in recent years generated a good deal of controversy. The claim was from the surviving relatives of a high school teacher who had committed suicide. His family claimed his death was work-related. The courts eventually sided with the family, agreeing that unusual job stress had contributed to his state of mind at the time of the suicide. The decision set a significant precedent for the workers' compensation system because of the high incidence of suicides in Japan. Increasingly, survivors of victims of other stress-related deaths are also seeking workers' compensation on the grounds that the deaths resulted from overwork. These efforts have been supported in recent years by the activities of the National Defense Council for Victims of Karōshi.

Industrial Organization for Safety and Health

The Industrial Safety and Health Law specifies an organizational structure companies of a designated size (generally fifty or more employees) must have so as to promote safety and health. It specifies that there must be a general safety and health director, a safety supervisor, a health supervisor, an industrial physician, licensed operation chiefs responsible for hazardous operations such as high-pressure rooms (used in underwater construction), and an overall safety and health controller where there are more than two employers in one work site. According to the directives, the manager who has overall control of the workplace (e.g., the factory manager) should assume the position of safety and health director. Specific functions may be delegated, however, to people such as the manager of the plant's safety and health department. This manager may fill the positions of safety supervisor and health supervisor concurrently, or other individuals may fill them and report to this manager.

Primary responsibility for safety and health flows from the general safety and health director to line managers in production according to the normal chain of command. In theory at least, line managers in many U.S. companies are also responsible for safety and health. Japanese line managers assume these responsibilities to a far greater extent, however. In Japan, safety and health staff act as technical advisers, provide safety and health instruction, and fulfill recordkeeping functions. They inspect for hazards and correct those they find. They also have primary responsibility for ensuring worker compliance with work rules.

Workers are relied on to a great extent in the Japanese safety and health system. Individual workers are held strictly responsible for following work rules and for avoiding accidents. This approach is consistent with the behavioral orientation of the Japanese system.

Workers have input regarding safety and health operations in Japanese companies through several avenues but primarily through a legally mandated company or plant safety and health committee (or a separate safety committee and health committee). These committees investigate the causes of accidents and illnesses, propose countermeasures to eliminate them, consider workers' complaints, and deliberate on other safety and health matters. Instead of committees, some firms use a joint consultation scheme. This approach, established by some union-management agreements, has been used since before the passage of the Industrial Safety and Health Law.

Membership on safety and health committees is determined jointly by labor and management. The chair of the committee is the general safety and health director or his or her representative (e.g., the manager of the safety and health department or the personnel manager). About half the members of the committee are appointed by the chair from lists of workers recommended by the labor union where one exists. If a majority of workers are not organized or there is no union, committee members must be nominated by a majority of workers.

Many companies also have safety and health committees at lower levels of the organization. And safety and health issues are often discussed in quality circles since safety and health are believed to be directly related to quality and productivity.

Yet another avenue for worker involvement in safety and health is through "toolbox meetings" on the plant floor. These short meetings occur within work groups each day before the start of work for the purpose of discussing safety and health issues. The meetings are intended to provide workers with an opportunity to raise safety and health concerns with their immediate supervisor. Occasionally, similar meetings are held after work to discuss safe commuting or other off-the-job safety issues. Toolbox meetings are voluntary in that they are not required by law, but there are strong social pressures for everyone in a work group to attend.

Many unions also have their own safety and health experts, and shop stewards typically have safety and health responsibilities in addition to their other duties. Unions also send individuals to training courses sponsored by the government and organizations such as the Japan Industrial Safety and Health Association to prepare them to serve as safety and health instructors at their plant.

Various companies involve employees in other government-encouraged safety and health procedures. These include training to foresee danger, safety checks, morning safety lectures and exercises, industrial safety and health suggestion systems, and safety inspections or patrols.

The basic goal underlying all of these activities is to increase employees' awareness of the importance of safe and healthful work practices. Increasing awareness is viewed as important as making actual improvements in the physical environment, again reflecting the greater emphasis the Japanese place on a behavioral orientation to avoiding injuries and illnesses.

Volunteerism in Occupational Safety and Health

There is considerable reliance on volunteerism in the industrial safety and health regulatory system in Japan, as reflected in the government's policy of providing "administrative guidance." Further, several voluntary organizations receive government support to promote occupational safety and health. The Japan Industrial Safety and Health Association (JISHA) is the most significant of these organizations. JISHA, originally named the Japan Industrial Safety Association, was established by government initiative through the Industrial Injury Prevention Organization Law of 1964 to encourage voluntary safety activities by industry. It is similar in certain respects to the National Safety Council in the United States, but its activities are more extensive, in part because of the government support. JISHA receives about 30 percent of its budget from the government and the remainder from member fees and sales of its publications and services (Japan Industrial Safety and Health Association 1985; Tomihisa and Yoshida 1986).[1] The National Safety Council, by contrast, is strictly a private organization.

JISHA provides a wide range of activities, including training programs and seminars on industrial safety and health practices. One component of this training is its Safety and Health Education Center, which trains workers who then provide safety and health training to fellow workers at their work sites. The center also publishes various training materials, including books, pamphlets, posters, and a periodical bulletin. Annual publications, such as the *General Guidebook on Industrial Safety* (published in conjunction with the Labor Standards Bureau) and the *Safety and Health Data Book in Japan*, provide comprehensive overviews of occupational safety and health experience and activities in Japan. JISHA also engages in research on industrial hazards and countermeasures, including studies on the safe use of video display terminals; research on preventing industrial accidents; research on hazards associated with the development and use of superconductors; studies on deep-sea diving and various services such as golf courses and security services; research on the use of agricultural chemicals; and studies on the prevention of occupational traffic accidents and backaches. It provides consultation services and safety and health subsidies for small and medium-sized businesses. In addition, JISHA sponsors safety contests and develops and promotes the use of safety and health slogans that appear on signs, ubiquitous throughout factories.

Several other voluntary organizations are concerned with occupational safety and health. These include associations in the construction, land transportation, forestry, mining, and cargo-handling industries. All of these associations as well as various nationwide employers' organizations and prefectural safety and health promotion organizations are members of JISHA.

Current Problems and Concerns

As noted, most Japanese who are directly involved with the occupational safety and health system in Japan are reluctant to discuss its problems. They are much more eager to talk about the many good features of the system. It takes prodding to get beyond "party-line" (*tatemae*) statements. Those outside the mainstream government-business-labor occupational safety and health coalition are more forthcoming in discussing the problems, and their comments were used to prod "insiders."

Current concerns, if not problems with the system, include psychological and physical stress on workers as a result of the fast work pace and the long work hours. The psychological stress can result in problems such as depression and alcoholism and, in extreme cases, suicide or other stress-related fatalities. The physical stress undoubtedly produces significant problems as well, especially cumulative trauma disorders (CTDs) and other injuries and illnesses. Most Japanese workers and managers are not willing to talk about the safety and health implications of the work pace for fear of sounding as though they are criticizing the Japanese work ethic. Complaining would entail a loss of face. Still, the long work hours are generally conceded to be a problem—although initial recognition seemed to have more to do with economic, political, and social factors than with concern about occupational safety and health—and the Labor Standards Bureau is supporting efforts to reduce the workweek. The Labor Standards Bureau has also begun paying more attention to CTDs in recent years, but, as we shall see, the level of concern is still nowhere near as high as in the United States.

Although individually many Japanese decline to complain publicly about the work pace or working hours, dissatisfaction is obvious. A recent survey by the Japan Institute of Labour (1989) found that 67.3 percent of salaried workers felt physically exhausted at the end of the day and 55 percent of all workers suffered from strong anxiety, worry, and stress. Similarly, a survey by the Federation of Japan Auto Workers' Unions found that 86 percent of workers complained about the pace of work and the amount of compulsory overtime (Solo 1990). Obviously, Japanese workers find it easier to complain anonymously than in person.

Yet another result of the work stress is a distinctly Japanese malady, holiday sickness, which manifests itself in both psychological and physical problems in workaholic employees who spend a weekend or vacation away from work. Some sufferers reportedly check into clinics to receive counseling and treatment for the physical manifestations of the problem, which disappear as soon as the individuals return to work (Helm and Gaffney 1986).

An even more serious problem is death from overwork, which was apparently common enough in recent years that the Japanese coined a term for it: *karōshi*. Yuzo Yamaguchi (1991) reports that in 1987 an estimated 150 Japanese died from overwork, although he acknowledges that establishing excessive overwork as a cause of death can be difficult. In addition to the National Defense Council for Victims of Karōshi, several organizations for the family members of survivors of *karōshi* have been formed in recent years to address problems related to excessive work. They seek to improve working conditions and help one another during the often difficult struggle to have the work-relatedness of the death recognized in order to establish eligibility for workers' compensation.

The high incidence of suicides is another safety and health concern in Japan. According to United Nations statistics, the suicide rate in Japan is more than one and one-half times higher than in the United States: 19.4 suicides per 100,000 in Japan compared with 12.1 per 100,000 in the United States (United Nations 1988). Determining whether a suicide was work-related often leads to confusion and controversy. The social acceptability of suicide, deriving from the long history among the Japanese to view it as an honorable way to deal with certain seemingly insoluble problems, further complicates the issue.

The aging of the work force is yet another source of concern with safety and health implications. The Japan Institute of Labour (1988) reports that the mean age of workers sustaining work-related injuries is increasing every year. It also reports that workers in their fifties incur injuries at twice the rate of workers in their twenties. Moreover, a number of specific illnesses (both work- and nonwork-related) associated with age, such as diabetes, heart disease, and cancer, are likely to become more prevalent with the aging of the work force, as are certain injuries, such as muscle strains and pulls.

Labor Standards Bureau officials note that several specific industrial groups are a particular concern regarding safety and health. They are construction firms, smaller firms, and subcontractor firms. Each of these groups has higher than average levels of injury and illness rates. Construction firms, which employ 10 percent of all workers in Japan, account for 42 percent of all fatalities and 32 percent of all injuries and illnesses; and establishments with fewer than one hundred employees account for 82 percent of all injuries and illnesses (Japan Institute of Labour 1988). Moreover, the Japan Industrial Safety and Health Association (*1988/1989 Annual Report*) notes that the disparity in injury and illness experiences between large and small firms is actually increasing. Part of the problem is that smaller firms are not required to follow the safety and health practices that larger firms must follow and apparently cannot be relied on to do so voluntarily.

Another perceived problem is the use of asbestos in the workplace. Critics contend that the required safety and health safeguards in Japan are less protective of workers than those in the United States. Another special area of concern that Japan and the United States share is the use of toxic chemicals in microchip manufacturing.

There are two additional personnel-related problems, although no one involved with safety and health programs in Japan acknowledged them. One is the relative lack of professionally trained or formally educated safety and health specialists, such as industrial hygienists and safety engineers. Most of the staff members of safety and health departments have a general engineering or science background rather than formal training in occupational safety and health. They, of course, receive training in safety and health once they are on the job, and they must pass government licensure tests to perform certain functions.

The second such problem is the relative lack of prestige accorded safety and health managers in Japanese corporations. Although it is difficult to measure, the safety and health function is clearly accorded lower prestige than line-management and most other staff functions. This is, of course, no different than in the United States. Given the Japanese seniority system, which involves transferring older workers so that they will not be supervised by a younger boss, the potential exists for safety and health departments to become dumping grounds for "deadwood" managers.

As in other industrialized countries, technological changes in Japan have had both beneficial and detrimental effects. The rate of advance in Japan has been so rapid, however, that both the benefits and detriments are especially noticeable. Technological innovations have enabled the Japanese to eliminate many dirty, hard, and hazardous jobs that can now be performed by machines. Overall safety and health conditions undoubtedly have improved as a result. But in some cases these machines have become hazards in themselves. For example, the Japan Institute of Labour (1988) reported that as of April 1988, industrial robots had caused twelve work fatalities. It also reported increases in complaints of eye fatigue and physical strains caused by the increased use of video display terminals.

Cooperation and Conflict

Occupational safety and health in Japan is described in written accounts (Japan Institute of Labour 1982 and 1988) and in person by involved parties as an issue of near-total cooperation among labor, management, and government. In sharp contrast to the situation in the United States, labor and management in Japan have nothing but good things to say about the occupational safety and health regulatory system. Each of these three parties seems quite satisfied with its working relationships with the other two parties, and, as with other areas of labor-management-government relations, a great deal of emphasis is placed on maintaining harmonious relations.

Although the predominant orientation toward safety and health in Japan is cooperative, several aspects of safety and health relations are conflict-oriented, reflecting broader conflicts in labor-management relations. One such conflict pertains to the reporting of injuries and illnesses. Japanese culture and industrial relations emphasize group loyalty and the obligations of group members. Being involved in a work accident (or incurring an occupational illness) is typically viewed as having let down one's group, especially in companies that emphasize the zero-accident campaign. One accident, injury, or illness means that one's work group and, in fact, the entire company fails to meet this goal. There is thus conflict at the group and individual level regarding injury and illness reporting. Reporting an incident brings attention to one's own failure and that of one's group. Not reporting undermines the injury and illness recording, reporting, and evaluation systems in the plant, the corporation, the industry, and the nation as a whole. Further, not reporting clearly conflicts with the *jidōka* concept emphasized in Japanese manufacturing management of bringing problems to the surface where they can be corrected.

The extent of the underreporting is difficult to measure. One Japanese researcher has speculated that the actual injury and illness rate is at least three times the reported rate (Fujimoto 1986). Based on the data analysis in chapter 6, the rate is probably higher than that. Nonreported injuries and illnesses are more likely to be minor, but I have heard reports of workers who attempted to conceal broken arms, broken legs, and concussions.

Managers also attempt to conceal injuries and illnesses, in some cases by having workers treated at the company hospital. Workers recovering from an injury or illness may also be assigned "rehabilitation work" so that they are not counted as having a lost-time injury or illness. Furthermore, there appears to be a pattern among workers and managers of simply not reporting cumulative trauma disorders, which are the fastest growing category of ailment in the United States. In 1988, for example, there were only 50 cases of "diseases due to vibration" in all of Japanese industry, although officials of the Labor Standards Bureau cited them as an increasing problem, and only 267 cases of "occupational back pain not caused by injuries" in the same year (Japan Industrial Safety and Health Association 1989).

There are other odd and suspicious recordkeeping conventions. One automotive company records, in a clearly mislabeled "total-injury" category, only those injuries and illnesses from which workers took six or more days to recover, regardless of whether they missed any work. Definitions governing inclusion in these statistics are obviously restrictive.

The incredibly high cost of building new facilities in Japan leads to numerous conflicts between labor and management over safety and health hazards. Thus, workers complain about dust, fumes, heat in the summer, and cold in the winter. Such conditions cause discomfort and frustration for workers and, in some cases, hazardous working conditions.

Another area of labor-management conflict pertains to the enforcement of such work safety and health rules as the wearing of protective equipment. Although safety and health managers frequently contend that work-rule compliance is complete, this is clearly a *tatemae* contention. In Japan, as in other countries, workers sometimes forget about long-term or low-probability hazards and seek immediate comfort or convenience.

Another set of conflicts pertains to the subcontractors and their relations with master employers. The relatively unfavorable financial positions imposed on subcontractor firms by master employers seem to result in the subcontractors cutting corners on safety and health. The subcontractors may also be assigned more hazardous work by master employers interested in achieving low injury and illness rates.

The description provided by Jared Taylor of conditions in subcontractors relative to master employers is a fairly typical Western account: "The two tiers are worlds apart in pay, prestige and working conditions. I have seen many second-tier shop floors that were greasy, rag-littered places with mobs of workers struggling over old, wheezing machines. First-tier shops are as clean as a hospital, with a few well-groomed workers to tend the gleaming machinery" (1983:149).

Because of the nature of this study and concerns of many Japanese that their safety and health system be seen in a favorable light, I was taken only to relatively large subcontractors with several hundred employees or more. The safety and health conditions there were noticeably inferior to those in master employers. Nonetheless, they were far superior to those in smaller subcontractors, which one can observe from the outside simply by walking down a city street. Other eyewitness accounts from Westerners who were not investigating occupational safety and health issues confirm that the conditions in some smaller subcontractors are very bad indeed. The Japan Institute of Labour (1982) has also reported that injury and illness rates are much higher than in master employers.

Partly in response to this situation, the government requests that master employers oversee the safety and health activities of their subcontractors and sponsor subcontractor associations to promote safety and health. This involvement with safety and health typically extends only to the first of many layers of subcontractors. Also, some firms distinguish between those subcontractors in which they hold an equity position and those in which they do not and take no responsibility for the latter. Lower-level subcontractors are left to manage their safety and health activities on their own. Neither the master employers nor the Labor Standards Bureau, with its limited number of safety and health experts, provides much assistance. Safety and health personnel who work for master employers are often reluctant to have any contact with their subcontractors for fear of being held responsible for improving their safety and health conditions. Furthermore, workers are likely to be more reluctant to complain to the Labor Standards Bureau about hazardous conditions when they lack adequate union protection.

The situation in subcontractors is exacerbated by the importance attached to within-group/out-of-group distinctions. Far less attention is paid to safety and health aspects of work affecting anyone outside the closely defined corporate family, even those having considerable interaction with it. Such parties do not benefit from the cooperative labor-management-government relations. Thus, when a visitor to a Japanese auto plant alerted a Japanese safety and health professional to an unsafe action by a worker on the plant floor, the response was, "Don't worry, he's not one of our workers; he's a supplier."

While inferior safety and health conditions are likely to be found in the smaller establishments of any industrialized country, the disparity appears more obvious in Japan. The exceptionally good safety and health conditions in master employers' facilities make the contrast even greater.

In addition to traditional safety and health activities, corporate safety and health departments in Japan are responsible for other functions that occasionally are sources of conflict. In one company, for example, where security is handled by the safety and health staff, they are responsible for convincing workers not to say anything negative about the company in public. In an intriguing and cryptic remark, safety and health officials of this company said that, for instance, they tell workers not to talk about the company's "spying activities."

As shown in figures 3-2 and 3-3, the cooperative versus assertive nature of management's behavior toward government and labor regarding safety and health can be plotted and compared with relations in the United States. Although these figures may oversimplify differences between industries and companies within countries, they nevertheless convey important information about tendencies in labor-management-government relations.

In the United States, management's behaviors toward both government and labor relative to safety and health can best be described as competitive; that is, there are high levels of assertiveness and low levels of cooperativeness. The behaviors of labor toward management and of government toward management are also generally competitive. The U.S. auto industry, which has had an agreement with the UAW since 1973 calling for labor-management cooperation on safety and health, is an exception. Management's behavior is close to the collaborative position, high in assertiveness and moderately high in cooperativeness. This position also describes labor's behavior toward management in the auto industry. As in the rest of the economy, however, the behavior of management and government toward each other remains competitive.[2]

In Japan, management's behavior toward government and labor can best be characterized as collaborative, that is, high in cooperativeness and moderately high in assertiveness. The behavior of labor and government toward management is also high in cooperativeness. Labor is

Figure 3-2. Character of Conflict-Coping Behavior by Management toward Government

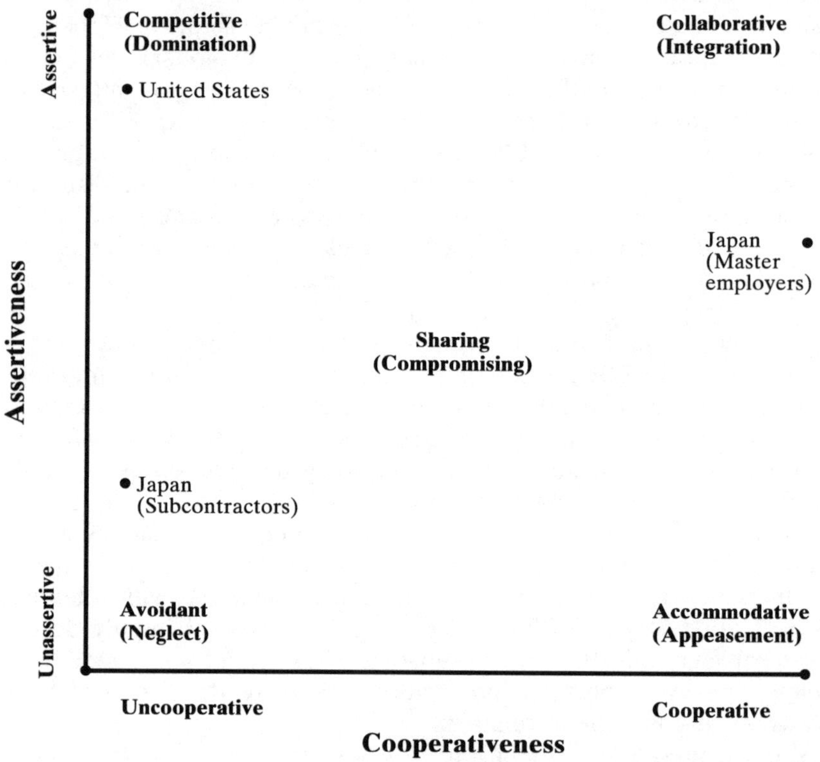

only moderately assertive toward management, however, and, because of the nonpunitive enforcement policies of safety and health regulators, government is only moderately assertive toward management. Thus, there is extensive cooperation among these parties. Management is assertive, but government and labor are progressively less so. Labor's position could be characterized as bordering on one of appeasement and government's as nearly collaborative.

The above description applies to master employers. In smaller firms, management would best be characterized as dominating labor and labor as appeasing management. Government could dominate these employers if it had the resources or inclination to scrutinize them. The current situation, however, might best be described as one of neglect or avoidance by both government and management.

In sum, there is much more cooperation in master employer firms in

Figure 3-3. Character of Conflict-Coping Behavior by Management toward Labor

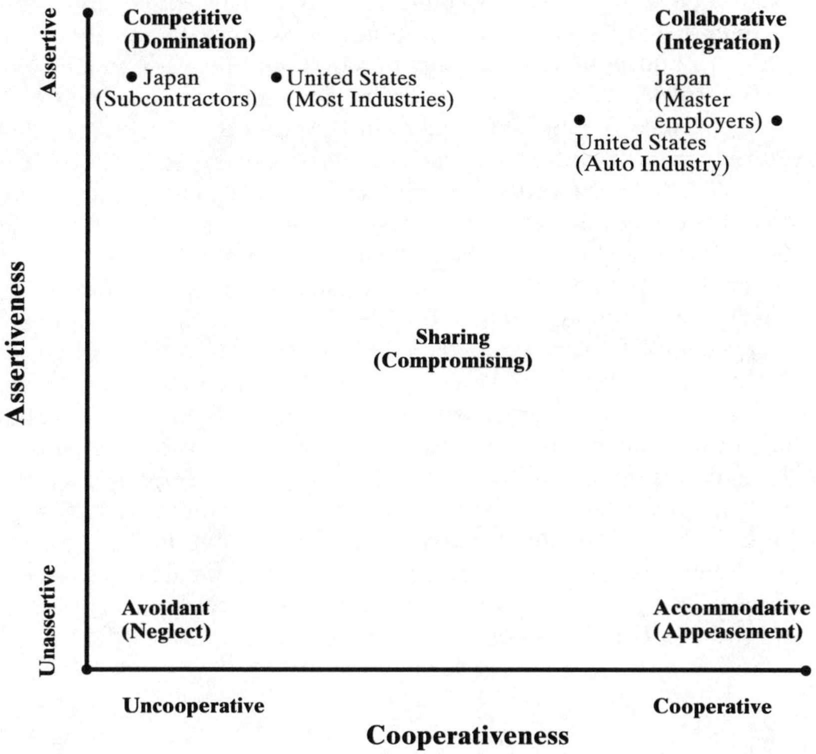

Japan with respect to safety and health than in U.S. industry in general, and labor and government are less assertive in their relations with management. The degree of labor-management cooperation is greater even than in the U.S. auto industry. Compared to master employers, however, there is far less cooperation among labor, management, and government in smaller Japanese firms, and labor and government are less assertive. I shall take up these issues again in chapters 4 and 5, where I consider safety and health activities at Jidosha in both its Japanese and American operations.

Enforcement Activities in Japan versus the United States

In the remainder of this chapter I shall compare the U.S. and Japanese occupational safety and health regulatory systems on several dimensions.

These dimensions include resources devoted to the regulation of occupational safety and health as measured by staffing levels and the activities themselves, including conducting inspections, citing violations, and issuing penalties. These dimensions tell us a great deal about the underlying philosophy and strategy of safety and health regulation in the two countries as well as how they are implemented.

There are considerable differences in the ways the two systems are structured, making simple comparisons impossible. These differences tell us a great deal about economic, social, legal, and political conditions in the United States and Japan and are related to differences in the missions of the agencies that have responsibilities for occupational safety and health. The Labor Standards Bureau in Japan has responsibility for monitoring a broad range of working conditions, including compliance with regulations pertaining to such matters as minimum wages, hours of work, and child labor. OSHA and state OSHA programs are more narrowly focused. This difference makes it difficult to compare directly those measures related to occupational safety and health activities. Further problems arise because of the unavailability of data on aspects of safety and health activities or because the results of these activities are unavailable in one country or the other. Nevertheless, the data tell us a great deal about the characteristic styles of occupational safety and health promotion in Japan and the United States.

Staffing

The broad focus of the Labor Standards Bureau makes comparing the numbers of inspections and inspectors difficult. The duties of the inspectors and the nature of inspections are also different in the two countries. In line with the broader mission of the Labor Standards Bureau, Japanese inspectors are responsible for inspecting more than just the safety and health conditions in the establishments they visit. Assisting Labor Standards Bureau inspectors are "expert officers in occupational safety and health." These officers lack the police powers of inspectors and typically act in a support capacity to inspectors. In short, neither Labor Standards Bureau inspectors nor expert officers in industrial safety and health are directly comparable to OSHA inspectors. Nonetheless, it is instructive to consider staffing levels and the numbers of inspections conducted for what they tell us about *changes* in occupational safety and health activities in Japan and the United States over time.

Table 3-1 contains data on the number of total personnel and inspection personnel at OSHA and at state occupational safety and health programs. As the data indicate, OSHA reached a peak level in overall

Table 3-1. Staffing of OSHA

	National OSHA		State Programs	Total
Fiscal Year	Total Positions[a]	Inspector Positions	Inspector Positions	Inspector Positions
1972	1,696	N.A.	N.A.	N.A.
1973	1,699	N.A.	N.A.	N.A.
1974	1,830	850	N.A.	N.A.
1975	2,435	1,350	N.A.	N.A.
1976	2,494	1,500	N.A.	N.A.
1977	2,717	1,600	1,139	2,739
1978	2,684	1,650	1,116	2,766
1979	2,886	1,828	1,132	2,960
1980	2,951	1,771	1,175	2,946
1981	2,655	1,686	1,110	2,796
1982	2,359	1,200	1,105	2,305
1983	2,284	1,125	1,081	2,206
1984	2,285	1,125	1,036	2,161
1985	2,239	1,200	N.A.	N.A.
1986	2,166	1,125	N.A.	N.A.
1987	2,211	1,125	N.A.	N.A.
1988	2,378	1,245	N.A.	N.A.
1989	2,441	1,277	N.A.	N.A.

Sources: U.S. Congress, Office of Technology Assessment 1985:222; OSHA Budget Office, unpublished statistics.

N.A. = not available.

[a]Total authorized positions before 1983; beginning in 1983, positions that are both authorized and funded are indicated, reflecting a recordkeeping change at OSHA.

staffing in 1980 with 2,951 total personnel positions and its peak for number of inspectors in 1979 with 1,828. The number of state-program inspectors reached a peak in 1980 with 1,175. The number of state and federal safety and health inspectors combined crested at 2,960 in 1979. Since 1980, there have been declines in all of these staffing levels. Total OSHA positions declined to 2,359 in 1982. Between 1982 and 1983, there was a change in OSHA recordkeeping, affecting comparisons of data from the 1974–82 period with data from the 1983–87 period. Thus, it is difficult to know what significance to attach to the apparent decline from 1982 to 1986 in the total number of positions. Clearly, there was an increase in the total number of personnel positions in the latter part of the 1980s following the apparent low in 1986 of 2,166. OSHA inspector positions declined to a low level of 1,125 in 1983, 1984, 1986, and 1987 but since then have increased to 1,277 in 1989. For the years data are available, the number of state program inspectors reached a low of 1,036 in 1984. This was the same year the total number of inspectors reached a low.

Japan has had no corresponding contraction in the staffing level of its occupational safety and health personnel. In fact, the number has increased steadily. The number of inspectors increased from 1972 to

1987 without any year-to-year decline during the period (see table 3-2). The number of expert officers in industrial safety and health also increased each year during this period, as did the total number of field staff (inspectors and expert officers combined).

There are considerably more Labor Standards Bureau inspectors than OSHA inspectors, although the U.S. economy is about twice the size of the Japanese economy. There were an average of 3,173 Labor Standards Bureau inspectors and 3,808 Labor Standards Bureau field staff for the years 1977 to 1984 (the only years for which staffing levels are available for both countries). For the same years, there were, on average, only 1,498 OSHA inspectors and 1,112 state program inspectors, for a combined average of 2,610.

T-tests for equality of means and F-tests for equality of variances were used to analyze the data. For the United States, the data used were the total number of OSHA personnel and the number of OSHA inspectors for the years 1974 to 1987. For Japan, the number of field staff were used for the same years. Looking first at total U.S. personnel, the results reveal a significantly higher mean ($t = 13.9631$, approximate $d.f. = 20.3$, $p. = .0001$) and a lower variance ($F = 3.25$, $d.f. = 13/13$, $p = .0424$) for Japan than for the United States.[3] The mean for Japan is also significantly higher ($t = 25.7848$, $d.f. = 25.0$, $p = .0000$) when one compares the number of U.S. inspectors and the number of Japanese field staff. There was no statistically significant difference in variances, however; the p-value is only slightly higher than the .05 criterion level ($F = 3.11$, $d.f. = 13/13$, $p = .0503$).[4]

The relevance of the differences in the absolute numbers is difficult to judge because of differences in staff responsibilities in the two countries. Labor Standards Bureau inspectors perform some functions that are carried out by personnel in OSHA as well as by personnel in other U.S. federal regulatory agencies. The existence of state-plan OSHAs in the United States further confounds comparisons. Differences in the variability of staffing levels for some measures and time periods are relevant and important, however. The steady growth in the number of field staff in Japan, for example, means that personnel resources are much more predictable than in the United States. This predictability undoubtedly facilitates planning and makes it possible for Labor Standards Bureau policies to be more consistent than at OSHA.

Inspections

Comparisons of the numbers of inspections in the United States and Japan provide some indication of the extent and coverage of regulatory

Table 3-2. Labor Standards Bureau Inspection Activities and Staffing

Fiscal Year[a]	Number of Inspections	Number of Inspectors	Number of Expert Officers in Industrial Safety and Health	Number of Field Staff[b]	Number of Inspections per Inspector	Number of Inspections per Field Staff Member
1972	250,538	2,940	279	3,219	85.2	77.8
1973	236,407	2,980	324	3,304	79.3	71.6
1974	219,122	3,010	384	3,394	72.8	64.6
1975	206,059	3,040	442	3,482	67.8	59.2
1976	173,744	3,070	501	3,571	56.6	48.7
1977	173,314	3,100	554	3,654	55.9	47.4
1978	176,359	3,130	605	3,735	56.3	47.2
1979	203,459	3,160	625	3,785	64.4	53.8
1980	204,910	3,178	639	3,817	64.5	53.7
1981	209,822	3,184	651	3,835	65.9	54.7
1982	208,108	3,202	664	3,866	65.0	53.8
1983	207,979	3,210	673	3,883	64.8	53.6
1984	210,607	3,216	676	3,892	65.5	54.1
1985	206,215	3,222	680	3,902	64.0	52.8
1986	202,767	3,230	682	3,912	62.8	51.8
1987	200,005	3,237	685	3,922	61.8	51.0
Average	205,588.4	3,131.8	566.5	3,698.3	65.6	55.6

Source: Labor Standards Bureau (Japan), unpublished data.
[a] Begins in April of the year noted and ends in March of the following year.
[b] Inspectors and expert officers in industrial safety and health combined.

efforts. These comparisons are not without problems, however, because of differences in the nature of inspections in the two countries, most notably that inspections in Japan cover numerous work issues. Table 3-2 shows the number of inspections carried out by the Labor Standards Bureau from 1972 to 1987. The number ranged from a high of 250,538 in 1972 to a low of 173,314 in 1977 and averaged 205,588.4.

Inspection data for the United States are available for both OSHA and state programs for 1973 to 1987 (see table 3-3). Excluding the start-up year of 1973, the number of inspections ranged from a low of 56,563 in 1981 to a high of 91,516 in 1976, and an average of 66,184 inspections were carried out each year. Data on state-program inspections are available for a more limited time period. For the years data are available, there was an average of 197,817 OSHA inspections and state-program inspections combined. This is about 4 percent lower than the average number of Labor Standards Bureau inspections during the 1972–87 time period.

The numbers of safety and health inspections in the United States (OSHA only) and Japan were compared using a t-test. The analysis showed that the mean number of inspections was significantly higher in Japan ($t = 26.7550$, d.f. $= 29$, $p = .0000$). This is particularly noteworthy given the size of the U.S. population and its gross national product relative to the size and GNP of Japan. There was no significant difference in variances ($F = 2.38$, d.f. $= 14$, $p = .1067$).

One measure of the efficiency of regulatory activities is the number of inspections conducted by each inspector. Undue emphasis should not be placed on this measure, however, since it also reflects the comprehensiveness of the inspection and the nature of the conditions inspected. Nevertheless, the differences are interesting.

Despite the more inclusive nature of inspections in Japan, each inspector carries out more inspections than in the United States (federal OSHA inspections only). During the period 1974–87, each Labor Standards Bureau inspector carried out an average of 66.0 inspections each year.[5] Each field staff member carried out an average of 53.3 inspections during this same period. During the 1974–87 period, each OSHA inspector performed an average of 52.25 inspections each year. Over the course of this period, the numbers of inspections per inspector at OSHA fluctuated greatly, from a high of 91.40 in 1974 to a low of 31.51 in 1979 and then to 63.69 in 1984. Inspectors in state OSHA programs performed an average of 97.50 inspections for the period 1979–84. This figure is considerably higher than figures for both OSHA and the Labor Standards Bureau for those years.

Two t-tests were conducted to compare the significance of mean dif-

Table 3-3. OSHA and State Program Inspection Activities and Staffing

	OSHA			State Programs		
Year[a]	Number of Inspections	Number of Inspectors	Inspections per Inspector	Number of Inspections	Number of Inspectors	Inspections per Inspector
1973	48,370	N.A.	N.A.	N.A.	N.A.	N.A.
1974	77,686	850	91.40	278,484	N.A.	N.A.
1975	81,728	1,350	60.54	176,898	N.A.	N.A.
1976	91,516	1,500	61.01	N.A.	N.A.	N.A.
July-Sept. 1976	18,279	N.A.	N.A.	N.A.	N.A.	N.A.
1977	59,227	1,600	37.02	N.A.	1,139	N.A.
1978	57,169	1,650	34.65	N.A.	1,116	N.A.
1979	57,603	1,828	31.51	107,636	1,132	95.08
1980	63,935	1,771	36.10	110,281	1,175	93.86
1981	56,563[b]	1,686	33.55	108,376	1,110	97.64
1982	61,686[c]	1,200	51.41	105,357	1,105	95.35
1983	67,727[d]	1,125	60.20	104,627	1,081	96.78
1984	71,652[e]	1,125	63.69	110,135	1,036	106.31
1985	71,719[f]	1,200	59.77	114,215	N.A.	N.A.
1986	64,335[g]	1,125	57.19	115,317	N.A.	N.A.
1987	60,117[h]	1,125	53.44	109,906	N.A.	N.A.
Average	66,184	1,367	52.25	131,021	1,112	97.50

Sources: U.S. Congress, Office of Technology Assessment 1985; U.S. President 1981, 1984, 1987, 1988; OSHA Public Information Office; OSHA Budget Office.
N.A. = not available.

[a]Years are fiscal years ending in the year noted. Until 1976, fiscal years began in July and ended in June. Beginning in 1976, fiscal years began in October and ended in September. The 1976 transition quarter is listed separately and is counted as one-fourth of a year in calculating the average number of inspections per year.

[b]Includes 17 records inspections.

[c]Inc!udes 8,663 records inspections.

[d]Includes 10,243 records inspections.

[e]Includes 9,270 records inspections.

[f]Includes 7,180 records inspections.

[g]Includes 4,631 records inspections.

[h]Includes 2,365 records inspections.

ferences and variance differences in the number of inspections per inspector in the United States (OSHA only) with the number of inspections per inspector and the number of inspections per field staff member in Japan. The mean number of inspections per inspector was significantly higher for Japan ($t = 2.4422$, approximate d.f. = 15.1, $p = .0274$); and the variance for the United States was higher ($F = 12.58$, d.f. = 13/13 $p = .0000$). The t-test, however, showed no significant difference in the means of inspections per inspector in the United States and inspections per field staff member in Japan ($t = .2331$, approximate d.f. = 14.9, $p = .8189$), but there was a significant difference in the variances ($F = 13.26$, d.f. = 13/13, $p = .0000$). The variance was higher for the United States.[6]

The greater number of inspections per inspector in Japan is important in light of the broader coverage of inspections there. Japan may be more efficient in conducting these occupational health and safety activities. The relative efficiency of the two regulatory systems merits further investigation but is beyond the scope of this study. The greater variance in the number of inspections per inspector in the United States compared with the number per field staff member in Japan is no doubt due to political influences on OSHA inspection policies (see Wokutch 1990a).

Violations and Penalties

Although there are important differences in the systems for issuing violations and charging penalties, a comparison of these aspects of safety and health regulation is instructive and relevant in describing the cooperative versus conflictual orientations of the two systems. Tables 3-4, 3-5, and 3-6 provide information on the enforcement activities of OSHA, state OSHA plans, and the Labor Standards Bureau. Table 3-4 shows the number of OSHA violations, the dollar amounts of OSHA penalties, the dollar amounts of penalties per violation, and the percentage of citations with contested claims. It also provides data on the number of violations and the proposed penalties (in dollars) set forth by state OSHA plans for several years.

The OSHA data, which are more comprehensive than the state-program data, show considerable variation in the number of violations (from a high of 382,073 in 1976 to a low of 97,906 in 1982), in the dollar amount of penalties (from a low of $3.7 million in 1973 to a high of $24.3 million in 1987), in the dollar amount of penalties per violation (from a low of $22.08 in 1974 to a high of $177.60 in 1987), and in the percentage of inspections with contested citations (from a high of 21.7 percent in 1979 to a low of 2.8 percent in 1985).

Table 3-4. OSHA and State Program Enforcement Activities

	OSHA				State Programs		
Year[a]	Number of Violations	Dollar Amount of Penalties	Dollar Penalties per Violation[b]	Inspections with Citations Contested (Percentages)	Number of Violations	Proposed Penalties (Dollar)	Dollar Penalties per Violation[b]
1973	144,527	$3,746,298	$25.95	5.2%	N.A.	N.A.	N.A.
1974	292,044	6,449,447	22.08	4.9	466,290	2,366,895	5.08
1975	316,386	7,043,601	22.26	6.0	494,160	6,526,881	13.21
1976	382,073	10,269,071	26.88	8.1	N.A.	N.A.	N.A.
1976 (July–Sept.)	39,300	2,406,715	34.73	10.7	N.A.	N.A.	N.A.
1977	178,118	8,201,720	46.05	12.5	N.A.	N.A.	N.A.
1978	134,406	12,736,373	94.76	18.2	N.A.	N.A.	N.A.
1979	127,466	13,861,071	108.74	21.7	275,442	8,167,912	29.65
1980	132,807	15,103,292	113.72	20.9	250,379	8,306,064	33.17
1981	110,489	7,099,763	64.26	10.8	234,738	7,072,818	30.13
1982	97,906	4,325,934	44.18	4.7	225,730	5,884,113	26.07
1983	108,862	4,817,807	44.25	3.2	226,244	5,235,746	23.14
1984	111,590	6,065,390	54.35	3.0	219,835[c]	6,517,537[c]	29.65
1985	120,024	7,756,647	64.53	2.8	265,435	8,309,314	31.31
1986	128,940	11,375,324	88.22	3.6	283,413	10,004,002	35.30
1987	136,751	24,286,596	177.60	4.6	284,652	11,189,399	39.31
Average[d]	167,980	$9,543,938	$67.71	9.2%	293,302	$7,234,607	26.91

Source: U.S. President 1981, 1984, 1987, 1988; unpublished OSHA data.
N.A. = not available.
[a]Years are fiscal years ending in the year noted. Until 1976, fiscal years began in July and ended in June. Beginning in 1976, fiscal years began in October and ended in September.
[b]Not all violations are assessed penalties.
[c]Includes four months of data from New York State. Subsequent years have data from New York for the entire year.
[d]The July to September quarter in 1976 is counted as one-fourth of a year in calculating the averages.

Table 3-5. Percentage of OSHA Initial Inspections Resulting in Citations and Penalties

Year[a]	Percentage Resulting in Citations	Percentage Resulting in Penalties
1972 (July–Dec.)	60%	46%
1973	67	51
1974	74	54
1975	79	57
1976	77	50
1977	70	30
1978	67	37
1979	65	38
1980	64	36
1981	60	29
1982	51	21
1983	52	23
Average	65.5	39.3

Source: OSHA unpublished data.
[a]Calendar years.

Data for the state programs are incomplete, but available data show considerable variation in the number of violations, proposed penalties, and dollar penalties per violation. In the most recent years for which data are available, there were increases in all three of these dimensions. The total number of violations cited by state programs has been more than double that of OSHA for most recent years, yet the dollar values per penalty have been much lower and the disparity on this dimension has been widening.

Table 3-5 shows the percentage of initial OSHA inspections in which there were citations and the percentage in which there were penalties. There is also great variability among these figures. In 1975, 79 percent of the initial inspections resulted in citations and 57 percent of the initial inspections resulted in penalties. Low figures of 51 percent and 21 percent respectively occurred in 1982.

Table 3-6 shows enforcement data for the Labor Standards Bureau. As with inspection data, enforcement data relate to a broader range of work conditions than just safety and health. The table shows the number of establishments with violations, the percentage of establishments inspected (number of inspections are from table 3–2) in which one or more violations were found, and the percentage of cases involving "sending the papers." In the years for which data are available, less than 1 percent of the total inspections resulted in sending the papers, reflecting the seriousness of taking this step.

The percentage of initial inspections in the United States that resulted

Table 3-6. Labor Standards Bureau Enforcement Activities

Year[a]	Number of Workplaces with Violations[b]	Percentage of Inspections with Violations	Number of Cases of Sending the Papers	Cases of Sending the Papers as a Percentage of Inspections
1972	N.A.	N.A.	1,393	.56%
1973	148,537	62.8%	1,575	.67
1974	138,282	63.1	1,395	.64
1975	130,043	63.1	1,363	.66
1976	108,413	62.4	1,773	1.02
1977	105,842	61.1	1,864	1.08
1978	104,976[c]	59.5	1,667	.95
1979	122,216[c]	60.1	1,545	.76
1980	124,871[c]	60.9	1,531	.75
1981	123,423[c]	58.8	1,426	.68
1982	125,781	60.4	1,443	.69
1983	124,729	60.0	1,413	.68
1984	N.A.	N.A.	N.A.	N.A.
1985	N.A.	N.A.	N.A.	N.A.
1986	N.A.	N.A.	N.A.	N.A.
1987	N.A.	N.A.	N.A.	N.A.
Average	123,374	61.1%	1,532	.76%

Source: Labor Standards Bureau (Japan) unpublished data.
N.A. = not available.
[a] Calendar year.
[b] Excludes violations concerning industrial work at home.
[c] Excludes numbers based on reinspections.

in citations was compared with the percentage of inspections in Japan that resulted in violations for the period 1973–83 using a t-test for equality of means and an F-test for equality of variances. The statistical analysis indicated that variance in the U.S. data was significantly higher ($F = 36.0$, d.f. = 10/10, $p = .0001$) but that there was no significant difference in means ($t = 1.7521$, approximate d.f. = 10.6, $p = .1088$). This indicates that, although there is no significant difference in the overall rate of citations per inspections in the United States compared with violations per inspections in Japan, there is a significant difference in the variation of these rates over time. The variation is much greater in the United States.

Although the differences in the patterns of issuing citations and violations are interesting, the differences in the imposition of penalties are far more significant. The charging of penalties is a far more serious and time-consuming process in Japan than in the United States and consequently is relatively rare. In the United States, the inspector may impose a penalty on finding a violation. In Japan, the inspector does not impose the penalty but rather recommends the case to the prosecutor's office for action by sending the papers. Moreover, the existence of a serious violation is not usually in itself adequate justification for sending the papers. The violation must either have resulted in an injury or illness or be for a condition that was formerly cited and not corrected. Other criteria for sending the papers are intriguingly described as "secret."

From 1972 to 1983, an average of 39.3 percent of initial inspections in the United States resulted in penalties. In contrast, in the same period, .76 percent of the inspections in Japan resulted in sending the papers. Furthermore, a much higher percentage of violations resulted in penalties in the United States than in Japan. In Japan the ratio of cases of sending the papers as a percentage of inspections to the percentage of inspections with violations was 1:80, whereas in the United States the ratio of percentage of initial inspections with penalty to percentage of initial inspections with citation was 1:1.7. Moreover, not all cases of sending the papers result in penalties. Based on the evidence, the prosecutor's office makes a determination whether to prosecute. A judge then decides whether a penalty should be imposed and, if so, the amount. Although no data on penalties were available, there is no doubt that the monetary value of penalties resulting from cases of sending the papers is minuscule compared with OSHA penalties in the United States.

A t-test revealed that for the years 1973–83 there was a significant difference in the means for the *percentage* of inspections with penalties in the United States and the *percentage* of inspections involving sending the papers in Japan ($t = 9.9860$, approximate d.f. $= 10$, $p = .0001$). The United States was higher, which is not surprising given the enormous disparity noted above. There was also a significant difference in variances ($F = 6{,}302.31$, d.f. $= 10/10$, $p = .0000$).

A t-test was used to determine whether there were significant differences in the variances and means for the *number* of violations in the United States (1973–87) and the *number* of cases of sending the papers in Japan (1972–83). The results showed that the two countries did differ significantly in their variances ($F = 9{,}999.99$, d.f. $= 15/11$, $p = .0001$) and in their means ($t = 7.8330$, approximate d.f. $= 15$, $p = .0001$). The United States had both a greater variance and a larger mean. Of course, all violations in the United States do not result in penalties, but neither do all cases of sending the papers in Japan.[7] Moreover, the sending-the-papers data include instances pertaining to matters other than safety and health. If these nonsafety and health transgressions were not included, the differences in means would be even greater.

To summarize, significant differences in the means exist for staffing levels, number of inspections, number of inspections per inspector, the percentage of inspections involving penalties or sending the papers, and the number of violations (cases of sending the papers in Japan). The means for the United States were lower for staffing, number of inspections, and number of inspections per inspector. They were higher for the United States for the number of violations and the percentage of inspections with penalties. Taken together, these comparisons show a distinct difference in national styles of safety and health regulation. The United States has a much more punitive approach, relying far more heavily on the threat of detection and the imposition of penalties to coerce compliance. In contrast, Japan relies much more on the willingness of employers to follow regulations voluntarily and to correct hazards once they are pointed out.

Dissimilarities in the responsibilities of inspectors, in the coverage of inspections, and in the meaning of citations and penalties pose questions regarding the meaningfulness of the differences in absolute numbers. Comparisons are meaningful, however, for examining the relative stability of regulatory activities over time. In these comparisons, the United States was found to have significantly higher variance in staffing levels (U.S. total personnel versus Japanese field staff), numbers of inspections per inspector, percentage of inspections resulting in citations or violations, the percentage of inspections with penalties or cases of sending the papers, and the number of violations or cases of sending the papers. In sum, the resources available to the Labor Standards Bureau and its enforcement activities are much more predictable than are the resources available to OSHA and its activities.

The lower degree of variability in staffing levels and enforcement activities no doubt can be attributed to the greater stability in Japanese politics, at least as they affect the regulation of occupational safety and health. Over the years several politically inspired changes in enforcement policy at OSHA tied to changes in the government and national politics have affected the enforcement of workplace safety and health regulations. There have been no comparable changes in Japan.

The higher status of Japanese civil servants compared with civil servants in the United States also affects the stability of safety and health enforcement. In neither country is occupational safety and health regulation viewed as a high-prestige governmental function. Nonetheless, the Labor Standards Bureau and the career civil servants who staff it are widely respected in Japan, whereas OSHA and its staff have been the frequent targets of scorn and ridicule. Because of its better reputation, the Labor Standards Bureau is more able to resist political pressure to change its policies and to resist pressures from employers and unions and their respective confederations.

In chapter 6 I shall assess the relative effectiveness of the U.S. and

Japanese approaches to safety and health regulation by comparing work injury and illness statistics for the two countries. Before that, however, we shift our attention to the *management* of worker safety and health. Examining worker safety and health at the firm and plant levels will shed further light on labor-management-government relations pertaining to occupational safety and health in the United States and Japan.

4

Occupational Safety and Health at Jidosha

This chapter describes occupational safety and health activities at a major Japanese automotive company, identified here as Jidosha.[1] Jidosha's safety and health activities are remarkably similar to those at other major Japanese auto firms, the six largest of which I studied in the course of conducting research for this book. The responses of safety and health managers to interview questions were so similar at these companies that by the last of my visits asking questions was merely a formality; the answers were obvious before the questions were asked.[2] The safety and health conditions and practices I observed were also similar. There is far less variation in the safety and health operations at these firms than there is among the Big Three auto firms in the United States. Where relevant, those differences in the activities of the Japanese companies are noted.

Although my focus is primarily the automotive industry, studies of major employers in other industries would probably yield similar findings, since major employers generally comply carefully with regulations and even with nonenforceable recommendations of the Labor Standards Bureau. Voluntary guidelines suggested by the quasi-governmental Japan Industrial Safety and Health Association are usually followed very carefully as well. In fact, government or JISHA publications were apparently the source of several responses to interview questions, which helps explain the similarities in responses.

The first part of this chapter provides background information on Jidosha. It then considers the structure and operations of safety and health activities, staffing levels of the safety and health function, and work injury and illness experiences at Jidosha and its subcontractors. Finally, it addresses cooperative and conflictual aspects of labor-management-government relations pertaining to safety and health.

Jidosha is a major automotive company that produces a full line of vehicles. Automotive products constitute the overwhelming majority of the company's sales; nonautomotive sales are less than 2 percent of the total. In the late 1980s Jidosha employed more than twenty-five thousand workers and produced more than 1 million vehicles each year. The output of forty-two vehicles per worker (including headquarters staff) in 1986 is high by U.S. standards but typical for Japanese automotive firms.

Jidosha has production and/or assembly operations in more than fifteen countries, the most important of which is the United States. The United States, after Japan, is also the most important national market for Jidosha's vehicles. Like most other Japanese auto companies, Jidosha has had various commercial relations and joint ventures with U.S. auto manufacturers for a number of years.

The overwhelming majority of Jidosha's operations are located at one giant production facility in a medium-sized Japanese city. Company headquarters are also at this site. Jidosha has operated there since it began production before World War II. As is common in Japan, severe space limitations have constrained expansion. As a result, the company recently built a new plant sixty miles from the main facility. At the time of its construction, many industry experts considered the new plant the best designed automotive plant in Japan for facilitating the flow of production. It also incorporated new production processes (at least new at the time for Jidosha) that, if judged successful, would be adopted at the main plant. These included removing the doors after painting so that workers could more easily install components in the interior and tilting each vehicle at various angles to give workers easier access to hard-to-reach areas, such as the underbody. This plant was the first Jidosha constructed at a site in Japan away from the main plant, although several plants had been built overseas.

Labor Practices

The labor practices at Jidosha are typical of those at large Japanese companies and as such differ from those at U.S. companies. Workers at Jidosha receive the benefits of lifetime employment, generous wages, extensive fringe benefits, and, in certain respects, pleasant working conditions. In return, workers demonstrate a high degree of loyalty to the firm and a strong commitment to work.

The wages of production workers differ from worker to worker. They are based 55 percent on age and 45 percent on ability and acquired skills. In the 1985–86 fiscal year, the average production worker received 3,240,000 yen in regular wages, 681,000 yen in overtime wages, and

1,304,000 yen in bonuses (half paid in July, half paid in December). At the exchange rate existing at the end of 1986 ($1 = 155 yen), this converted to a nominal yearly wage of $33,710, roughly comparable to that earned by UAW production workers in any of the Big Three auto companies in the United States. This comparison may overstate the purchasing power of the Japanese worker, however, in that the exchange rate in 1984 was $1 = 250 yen. In other words, using that exchange rate, 1985 wages would have converted to $20,900. The strengthening of the yen accounted for very little increase in the standard of living for the typical worker since prices, set in yen, did not decline much, if at all, even for imported goods. Wages are more accurate and significant for evaluating the labor component of manufacturing costs since a large percentage of Jidosha's output is sold in the United States for U.S. dollars.

Jidosha workers receive a wide range of benefits from the company. These include government-mandated health insurance, use of a 350-bed company hospital that provides subsidized medical care to workers and their families (as well as to members of the local community on a for-profit basis), inexpensive *bentō* (box lunches served to production workers in cafeterias and to office workers at their desks), a company store that provides a wide range of goods at discounted prices, discounts on the purchase of cars, and low-interest loans for employees purchasing or building a house. The company also provides residences for newly married employees and dormitories for single workers living away from home. Other benefits, such as company-sponsored vacations, training, and retirement plans, are also provided.

The normal work schedule at Jidosha is officially described as "an eight-hour day, forty-hour week." As with other Japanese employers, there is sensitivity at Jidosha about the length of the workweek because of the national and international pressure on Japanese employers to reduce working hours. Descriptions of the workweek must therefore be viewed in the context of *tatemae* communication. Workers work some six-day weeks; eleven Saturdays are considered "normal," and workers do not receive the 40 percent premium paid for "overtime" for these days. Some of these "normal" Saturdays are reportedly negotiated in exchange for extended vacations at the time of company and national holidays throughout the year. In the 1985–86 fiscal year there were 250 scheduled workdays, including thirteen Saturdays. Saturday work permits extra production, as does the one to two hours of overtime typically worked each day, for which workers are paid a 30 percent hourly premium. Weekend work also enables the company to take advantage of cheaper utility rates. In general, overtime work is viewed as cheaper and more flexible than investing in new production technology or hiring more workers.

In addition to fifteen national holidays and various company holidays, workers receive fourteen to twenty personal vacation days, depending on their length of service. Workers at Jidosha and at other Japanese auto firms use relatively few of these holidays, however. In 1984–85, the average production worker worked 2,349.2 hours. This is well in excess of the hours of work of production workers in typical U.S. or European automotive factories, who generally work no more than 2,000 hours per year and could work as few as 1,800 to 1,900 hours on a full production schedule.

Approximately 6 percent of Jidosha's twenty-five thousand–plus workers are women. There are no women on the production line, however, in the Japanese facilities of Jidosha. In fact, the only female production workers I saw in any of the six Japanese automotive firms I visited were in the motorcycle assembly plant of another automotive firm.

Jidosha employs more than 250 handicapped workers. The government requests or recommends that large companies like Jidosha employ the handicapped (Mafune 1986). The number employed at Jidosha corresponds with the number recommended (about 1 percent of total employment) for a company of Jidosha's size.

Total absenteeism was cited as 3.3 percent for the 1984–85 fiscal year. This figure includes people who were taking personal vacation days (but not company holidays), sickness and injury leave, bereavement leave, marriage leave, maternity leave, personal leave, leave in preparation for a transfer, and menstrual leave. Most were taking personal vacation days or other contractually negotiated days off. Controllable absenteeism was only 0.3 percent.[3] Rates in previous years were approximately the same.

Most workers do not take full advantage of their personal holidays, which they may use for actual vacations or when they have nonwork-related illnesses or injuries. In the 1984–85 fiscal year, workers used, on average, only 40 percent of their personal vacation days. This low level is significant in that there are no direct financial incentives to work on these days.[4] Further, workers may accumulate only forty vacation days. Even if a worker uses all of his or her personal vacation days and needs additional days because of nonwork-related injuries or illnesses, he or she is still compensated through the Jidosha Health Insurance Program at a rate of 80 percent of his or her pay. (This exceeds the 60 percent compensation level mandated by the government.) The company pays about 65 percent of the premium for the insurance program; the worker pays the difference.

The company and union policy is to encourage workers to take advantage of their personal holidays as a way to improve their overall health. In addition, the company encourages supervisors, co-workers, and members of the safety and health staff to be on the lookout for employees who are working in spite of an infirmity. The morning safety and health check is one way of doing this. Nonetheless, some workers indicated that foremen let them know quite clearly that they are needed on the job and that taking vacation days except when absolutely necessary for health reasons would not be appreciated. Moreover, because there are few redundant workers in Jidosha plants, absences are likely to create work for fellow work-group members and/or the supervisor, leading to resentment. Given the rapid pace of work when there is full attendance, the social pressure not to take off can be strong.

At times, the attitudes toward absenteeism seem almost laughable. In one group interview, a foreman, denying that workers feel pressured to work when they are sick or injured, unintentionally affirmed precisely what he attempted to refute. He contended that he would be willing to take off from work if he was sick or injured and in fact had done so. When questioned further, however, he said that he had missed one day of work, ten years earlier, because of severe back pain, which had prevented him from getting out of bed.

The increased use of personal holidays by young people in Japan is decried by some Japanese as indicative of a declining work ethic among the young. And Jidosha officials speak with mixed emotions when they note that the increased use of personal vacation days has led to an increase in the total rate of absenteeism in recent years. One gets the impression that they view this increase in the use of vacation days as evidence of a lack of dedication.

The turnover rate for Jidosha's Japanese operations was 2.38 percent in 1984–85. This is quite low by U.S. standards. A total of 679 workers quit, 66 at the mandatory retirement age of sixty. (As in other Japanese firms, the mandatory retirement age at Jidosha has been creeping up in recent years from the traditional age of fifty-five.) Thus, 613 workers quit for other reasons. No doubt, many were women who quit to get married. No one was discharged that year, and labor relations personnel said that discharging workers is very rare. In fact, they initially contended that no one had *ever* been discharged from Jidosha. I expressed incredulity given the size of the company and asked what was done if someone "went crazy or was found to have committed a crime." The initial response was that Jidosha did not hire people who had the potential to "go crazy or to commit crimes." When company officials saw this statement quoted in a report to management, however, they conceded that such problems do develop on rare occasions and that a disciplinary committee reviews the case and suspension or dismissal of the worker might result.

When someone's performance is found to be seriously deficient, supervisors and co-workers pitch in to bring the worker up to speed, just as fellow classmates do with slow learners in school. If the worker's performance remains deficient, an effort is made to find him or her a more suitable job. Some master employers send such workers to subcontractors, but Jidosha officials contend that they do not follow this practice. They believe that once someone is hired and becomes a part of the Jidosha family, management has an obligation to help that worker achieve his or her potential in the company.

The low turnover rate can be attributed in large part to the lifetime employment system. When the company underwent a serious downturn in the 1970s following the first oil crisis, several thousand employees were sent on "loan" to subcontractors and subsidiaries for several years, enabling the company to avoid layoffs. Many were sent to Jidosha dealerships, where they took part in an aggressive sales effort, including selling autos door to door.

Although Jidosha's commitment to lifetime employment was widely praised in this case, the practice has its down side. It is as though the workers have "golden handcuffs." Jidosha is probably the best-paying employer in the region, and most Jidosha workers clearly would like to finish out their careers with the company, but their commitment goes beyond that. Workers who cannot "cut it" at Jidosha would most likely be shunned by other major firms. Moreover, a large percentage of the smaller firms in the region are in some way tied economically to Jidosha. A worker who left Jidosha on unpleasant terms would have a hard time getting hired by them. Thus, the dedication and obedience shown by Jidosha workers are to some extent necessitated by economic self-preservation.

In addition to regular Jidosha workers, as of 1986 the company employed approximately 1,400 temporary full-time workers. Jidosha has hired temporary workers, at least in large numbers, only since 1984. They are not generally considered part of the corporate family and are often mentioned only as an afterthought in discussions of labor practices. Like other major Japanese employers, Jidosha hired as temporary workers many individuals who were displaced because of the privatization of Japan National Railways (JNR).

The base pay of temporary workers is about two-thirds that of regular workers. Specifically, it is 913 yen an hour, or 150,000 yen a month, compared with 230,000 yen a month for a permanent worker with the same skills, experience, and job (1985–86 figures). With overtime, temporary workers earn about 220,000 yen a month (about $17,000 a year), based on exchange rates at the end of 1986.

Temporary workers also receive different benefits. The most significant difference is that they do not receive semi-annual bonuses, which amount to approximately five months' wages. Further, they are not entitled to personal vacation days, not eligible for company-subsidized loans, nor able to participate in the company retirement program. Their contracts are for one year or less, and the company has no commitment to rehire them. Because of their limited experience with temporary workers, Jidosha officials are not confident of all the statistics, but they estimate that about 30 percent quit before their contract expires and that about half of the remaining 70 percent reapply. Statistics on temporary workers are not included in the above turnover rates but are included in injury and illness rates reported below.

Because temporary workers are not as well trained, experienced, or motivated as other workers, management expresses special concern about them with respect to occupational safety and health. One manager observed that "they have no sense of commitment." Obviously the company has little commitment to them either. As a result of these conditions, temporary workers sometimes pose a safety and health threat to themselves as well as other workers.

Like other Japanese employers, Jidosha attempts to improve its performance through group-oriented *kaizen* activities. Work teams, which meet in quality circles before or after work, systematically analyze aspects of the production process. Their goal is to improve productivity through "rationalization." According to one Jidosha official, "Elimination of waste is the basic principle: wasteful time, wasteful space, or wasteful something." Even minor improvements are welcome since even tiny changes in sufficient numbers can result in major improvements over time. Thus, company officials point with pride to the more than 2.2 million suggestions made by quality circles in one recent year.

Likewise, Jidosha employs a job-rotation system, also referred to as the multiskilled worker program. The goal of the program is to increase management's flexibility in being able to reassign workers. The program also helps promote an important long-term goal of *kaizen* activities, which is to reduce the number of workers required to perform a given set of tasks. Job rotation keeps workers more attentive to the job, thereby enhancing quality and reducing the likelihood of an accident. It also reduces the cumulative stress on specific body parts insofar as different jobs stress different muscles and bones. Not all workers or jobs are currently included in the job-rotation system, but the aim is to increase the number who are.

Consistent with the philosophy that "you can't inspect quality into a vehicle" at the end of an assembly line, individual workers are responsible for inspecting the results of earlier work during the entire production process. Signal buttons every twenty meters enable workers to notify their superiors if something goes wrong. One button activates a yellow light that immediately brings a supervisor to the work station. If he can correct the problem himself, he will, or he will stop the assembly line and seek further help. In the case of a serious problem such as an accident, the worker can activate a red light and immediately stop the line. A number of people will rush to the scene to assist. This system is consistent with the Japanese concept of *jidōka* (bringing problems to the surface where they can be addressed).

Organization of the Safety and Health Function

According to the organization chart and formal written and oral descriptions of the safety and health function at Jidosha, the president of the company has ultimate responsibility for worker safety and health.[5] This policy conforms with the dictates of Japanese safety and health regulations. In practice, though, responsibility is to a large degree delegated to the corporate industrial relations manager, who has general oversight responsibility for safety and health and other personnel-related functions. From the corporate industrial relations manager flow several lines of responsibility: line responsibility within plants and major operating units, staff responsibilities within these components, staff responsibilities at the corporate level, and committee responsibilities at various levels within the organization.

Principal day-to-day responsibility for safety and health at Jidosha is delegated to line managers, that is, plant managers and managers of major units (e.g., research and development, industrial machinery). From here, responsibility flows to division managers, department managers, superintendents, foremen, and finally to the employees, who are organized into work groups. Work groups have responsibilities for general housekeeping and preventive maintenance in their areas.

Although there are differences in the degree to which managers embrace their safety and health responsibilities, on the whole, they take them seriously. Many said that they felt it was their responsibility or fault when an accident occurred in their area, even if it occurred because of an employee's carelessness. These managers expected to be reprimanded by superiors when an accident occurred, and they felt such a reprimand was appropriate. They contended that the manager should have trained the employee to be less careless. Other managers were not as magnanimous, contending that such workers deserved to be blamed and scolded. One manager encompassed both of these positions:

> If this accident was found [to be] due to the worker's carelessness, [the] manager should scold him. . . . The line manager should share the responsibility and depending on if that accident is very big, the line manager should take full responsibility for the accident by changing working conditions or changing work facilities. If it was a small accident as a result of a worker's carelessness, the line manager should caution the worker to avoid the same accident.

Another manager in an engine plant stated, in ascribing responsibility for a recent accident, that it was 70 percent the fault of the worker and 30 percent the fault of management. In response to a further question, he conceded that the injured worker had indeed brought dishonor to his work group and that fellow workers would blame him for this.

The contention that managers bear sole responsibility for all accidents might best be understood as a *tatemae* position. It is consistent, however, with the Japanese tradition that the person in charge takes responsibility (at least symbolically) when something goes wrong. This tradition is exemplified by the practices followed after a fatality.

The standard procedure in Japan when a work fatality occurs is for a senior corporate official to call on the family of the victim to express sympathy on behalf of the company. Usually, that official, the company president, or some other high-level corporate representative also delivers a eulogy at the worker's funeral. Typically, the official presents a monetary gift to the family at the time of the visit. In 1986, Jidosha gave families of fatality victims 20 million yen (approximately $133,000) if the worker was married and 17 million yen if he or she was single (approximately $113,000). This support is provided in addition to government-mandated workers' compensation.[6]

This procedure was described to me by a prominent Japanese labor lawyer who handles many workers' compensation cases, although apparently none for Jidosha. He was critical of the size of the settlements, noting that automobile fatality victims typically receive about twice the compensation of work fatality victims. He also spoke somewhat cynically about the custom of delivering an apology to the family. He noted that his brother is a senior safety and health official and has been required to pay such visits. As a result, he refers to himself cynically as "the manager in charge of condolences."

Production workers expressed the view that minor accidents are their own fault, and some confessed that they are embarrassed to report injuries resulting from their carelessness. Although they said they conceal only minor injuries, they recounted incidents in which co-workers hid or attempted to hide major injuries such as broken limbs, concussions, and serious lacerations. Workers conceal injuries and illnesses for a number of reasons: concern that reporting the injury or illness will hurt production (if the worker has to miss work for treatment, for example), fear of being reprimanded by superiors, fear of being ridiculed by co-workers, and a desire not to let down fellow workers and superiors in the pursuit of the goal of zero accidents and zero diseases. Another contributing reason is embodied in the claim made by several line managers as well as safety and health staff that virtually all correctable hazards have been eliminated and that additional improvements in workplace safety and health can only come from changes in employee attitudes and behavior. This attitude would seem to reinforce the view among workers that any injuries or illnesses that do occur are their own fault and that they *should* be ashamed.

Management contends that workers hid injuries and illnesses only rarely in the past and that efforts to encourage workers to report injuries and illnesses have eliminated this problem. Given the comments of workers themselves, however, it is difficult to believe this assertion. In addition, chapter 6 provides data that indicate that there is substantial underrecording of injuries and illnesses in Japan.

Managers are unclear whether they are specifically evaluated on the safety and health performance of their areas. One manager's response to a question about this matter was unusually candid but representative: "Honestly, I'm not sure." Another pointed out the difficulty of evaluating performance on this dimension, contending that since zero accidents is considered the natural level, one would have to evaluate something other than injury (or illness) experience. There are no specific items relating to safety and health on performance appraisal forms; however, some foremen and managers felt that performance in safety and health could affect the evaluation on other dimensions. They added, however, that this would probably be the case only if safety and health performance was bad. They did not feel that good performance would affect an overall evaluation. Other managers seemed surprised or puzzled by the notion of evaluating safety and health performance. One department manager noted that there were only sixty line managers out of a total of fourteen hundred management-level personnel at Jidosha. Inclusion of safety and health performance, in his view, would almost certainly work to the line managers' disadvantage. He felt that few would want to become line managers if safety and health performance were evaluated. His comment is surely an exaggeration, however, given the relatively high status of line managers.

Corporate- and plant-level safety and health staffs operate in conjunction with line managers on safety and health activities. At the corporate level, the general safety and health supervisor is responsibile for the Safety and Work Environment Division and the Health Care Center Division; he reports directly to the manager of industrial relations. The Safety and Work Environment Division is further divided into the Safety Department and the Work and Hygiene Department, each of which is headed by a department manager. Including managers, there are a total of sixty-five staff members in these units: eighteen in the Safety Department, ten in the Work and Hygiene Department, thirty-five in the Health Care Center, and two general managers. As we shall see, this level of staffing is very high compared with U.S. auto companies.[7]

Two related units are the Jidosha hospital and a company fire station. The hospital is said to be responsible for "curing" injuries and illnesses, in contrast to the health care center, which is mainly concerned with prevention. Both the hospital and the fire station provide service to the public at large as part of Jidosha's contribution to general social welfare. The hospital is operated as a profit center—although Jidosha officials contend that it would be supported by the company even if it were operating at a loss. About 60 percent of its 850 daily visitors are members of the general public. The remaining 40 percent are Jidosha employees and their families.

The physical layout of the areas where the safety and health staff work is consistent with and illustrative of several features of Japanese management. Staff of the Safety Department and of the Work and Hygiene Department are in one section of a very large room that houses the entire Industrial Relations Department. Safety staff and Work and Hygiene staff are seated in two long rows, each of which consists of pairs of desks that face each other. Seating is according to seniority; the most senior staff are closest to the department manager. His desk is perpendicular to the row so that he can see all his staff. Needless to say, such an arrangement facilitates group activity. A manager can call a meeting simply by raising his voice. The efficiency of this arrangement should be apparent to anyone who has ever tried to schedule a business meeting for three or more busy people. This layout is typical of the office arrangement in other parts of Jidosha and at other companies in Japan.

Another group responsible for promoting occupational safety and health is made up of full-time plant-level safety and health experts. They are assigned to a division and have responsibilities similar to those of the corporate safety and health staff. These include inspecting work sites for hazards, training, recordkeeping, and accident/injury and illness reporting. They have principal responsibility for safety and health during night shifts, when department and division managers are not there. Plant-level safety and health experts closely coordinate their work with corporate safety and health staff as well as with line managers.

Yet another actor in the safety and health system at Jidosha is a subsidiary that, among its several lines of business, conducts safety inspections on machinery for Jidosha and other firms. Such an arrangement would be unusual in the United States, but it is not at all uncommon in Japan, where master employers and their subcontractors comprise a sort of extended family.

Staffing Levels

Safety and health staffing levels at Jidosha as a whole and at the main plant are very high in comparison with U.S. auto companies but not at all unusual by Japanese standards. There are more than 40 plant-level safety and health personnel at the main plant, resulting in a ratio of 1 plant-level safety and health manager for every 575 workers. This figure understates the staffing level, however, in that Jidosha's corporate headquarters and principal production facilities are in the same complex and many staff at the corporate level perform the same functions as plant-level staff in other companies. Looked at this way, there is 1 staff member for every 260 employees (more than 110 full-time staff for more than 25,000 employees). At another Japanese auto firm, the ratio was 1 safety and health staffer for every 109 workers. By comparison, a typical plant of a Big Three auto firm has 1 safety and health specialist for every 1,750 workers. Some of the difference may be because job descriptions in Japanese companies are more flexible than in American companies and even than in American subsidiaries of Japanese companies. For example, it is often the case in Japan that no one in a department is assigned exclusively to secretarial duties. At Jidosha, most of the workers included in the above totals would perform some tasks that would ordinarily be performed by a secretary in the United States.

Goals and Activities

The mission of the Safety and Work Environment Division and of the Health Care Center Division is to prevent accidents, injuries, and illnesses while at the same time promoting the general health and safety of employees. The Safety Department of the Safety and Work Environment Division is responsible for preventing accidents and resulting injuries. The Work and Hygiene Department is responsible for preventing illnesses resulting from excessive exposure to heat, noise, dust, and hazardous materials. The Health Care Center Division is responsible for monitoring and promoting the health of workers.

As is common in Japanese companies, Jidosha promotes safety and health consciousness through the use of various slogans. Thus, the Jidosha safety and health goal or credo is to "infuse the workplace with vitality and secure the safety and health of employees in the Jidosha group by respecting human beings under the slogan of 'There can be no production without safety' " (Jidosha 1986:2). This philosophy is reflected and elaborated on in more specific goals and objectives, safety and health activities, and other slogans.

Since the early 1980s, the company has participated in national zero-accident campaigns. These five-year campaigns were first promoted by the Japan Industrial Safety and Health Association in 1973. Safety and health objectives at Jidosha are explicitly tied in to the zero-accident campaign and are stated as follows: "Achieving zero accident and zero disease as a final objective by promoting the zero-accident campaign, reaching the lowest accident and disease level in the auto industry, and continuing the improvement of the work environment" (Jidosha 1986:2).

More important than providing operational goals, which may not be attainable in a company the size of Jidosha, the zero-accident campaign reflects a belief based on respect for workers as human beings that no accidents are acceptable. Thus, specific limits or targets are set at ambitiously low levels, although still above zero. For fiscal year 1986, these were .30 or fewer leave accidents per million hours worked (a total of 15 or fewer accidents); .40 or fewer nonleave accidents per million hours worked (a total of 20 or fewer); .05 or fewer lost workdays per thousand hours worked (a total of 3,000 or fewer lost days); 25 or fewer commuting accidents (leave and nonleave combined); and zero fatalities or major accidents.

In addition, several "practical principles" are set out yearly. For the 1986 fiscal year, these were threefold:

(1) Emphasis is placed on the safety control for exterminating vital accidents and after-effect accidents (catching by machine, traffic accidents).
(2) Labor sanitation [health] is fixed in the zero accident campaign while improving workshop environment.
(3) Companies in the Jidosha Group (subsidiaries, suppliers, contractors) are instructed and assisted (Jidosha 1986:3).

There are also specific targets in the area of health. For example, in 1989, all 209 workshops with noise levels above 85 decibels (db) were targeted by the Safety and Work Environment Division for improvement to the 85-db level over a three-year-period and 69 were targeted for improvement in the first year.

Jidosha also uses slogans as part of its zero-accident campaign. One that was used widely is "*Anzen daiichi*" or "Safety first." It sounds familiar since it was adopted from the United States. According to a JISHA publication (Tanabe 1984), the slogan was developed around 1900 by the president of the U.S. Steel Corporation, Elbert Gary. It was widely adopted by U.S. industry but was mainly used as a motivational tool for workers. JISHA suggests that to be truly effective, it must guide managers' and workers' decisions. The slogan "There can be no production without safety" (*Anzen nakushite seisan nashi*), which Jidosha also uses, is derived from JISHA's zero-accident campaign. The message is that safety is intricately tied up with the production process and that accidents are an extreme form of inefficiency that must be eliminated. Another way this message is stated is through the slogan "Safety, productivity and quality are all part of the same operating philosophy," which suggests that problems on one of these dimensions will hurt performance on the other two. All of these slogans are said to reflect an underlying respect for the worker as a human being.

The above slogans are somewhat abstract or philosophical in nature, but one is very concrete and pragmatic: "If something abnormal happens in the workshop, stop the machine, call the leader and wait." Attempts by individual production workers to repair malfunctioning machines are notorious for causing injuries; this slogan is used to discourage such behavior.

Slogans are also used to encourage worker safety and health off the job. For example, during the time of my visit the slogan to encourage safe commuting was, "Patiently, generously and tolerantly wait." Anyone who has braved Japanese rush-hour traffic knows that, unfortunately, few adhere to this maxim.

Much of the sloganeering and several of the safety and health activities I discuss below have strong ritualistic components. Perhaps the most distinct ritualistic practice is the "safety and sanitation [health] prayer." In keeping with the Shinto tradition of visiting shrines to ask the gods for help with life's many hurdles and challenges, Jidosha safety and health managers, along with officials of the prefectural office of the Labor Standards Bureau, visit one of the popular Shinto shrines at the beginning of each year to pray for the health and safety of employees. The safety and health staff were mildly embarrassed about discussing this practice, and their initial reaction to questions about it was laughter. In response to further questions about the laughter, one staff member said, "We are trying very hard to improve our work environment, but sometimes we are relying on God." He also asked, "Do American managers go to church for safety and health in their company?" (My response was that perhaps individuals do but that groups do not.) Company officials, after reviewing an earlier version of the manuscript for this book, offered a further explanation of the Japanese tradition:

> If readers have little knowledge about things Japanese, they may have grave misunderstandings when they read this. The safety and health prayer is a traditional event in Japan just like a rich harvest prayer. The prayer alone does not promote the "zero-accident campaign" but offers people an opportunity to pay attention to safety and health.

Jidosha undertakes an elaborate system of activities to promote safety and health throughout the organization. These activities are intended as much to promote worker conscientiousness about safety and health as they are to improve the physical conditions of the work environment. In this sense they reflect the behavioral orientation of Japanese worker protection.

Several of the management practices that have been credited in the literature with giving Japanese companies competitive advantages are applied as earnestly to safety and health activities. Thus, management is willing to undertake activities or investments that promise long-term returns without insisting on immediate results. In addition, participative decision making is practiced in the safety and health arena. As in other applications of this practice, it often slows down decision making, but it ensures greater commitment to a decision once it is made.

Kaizen activities are also applied to safety and health aspects of production as workers, managers, and safety and health staff constantly look for incremental ways to improve safety and health. Jidosha managers claim that *kaizen* activities aimed at improving safety and health are highly consistent with those directed at productivity and quality, reflecting the view that safety, quality, and productivity are integrated components of Jidosha's operating philosophy. As noted above, the job-rotation system is said to affect productivity and quality positively, as well as safety and health. My own observations support this assertion, although, as we shall see, conflicts also emerge.

In the following few pages, I describe six categories of safety and health activities. These categories are training programs, safety and health committee meetings, inspections, signs, hazard-prediction and prevention activities, and accident reports and investigations. These activities are ranked according to their importance as perceived by safety and health managers. Several other safety and health activities are also described.

Broadly speaking, safety and health are promoted through three approaches: improving machinery and facilities, improving administration, and improving work practices of individual workers. Each of these approaches is reflected in specific activities described below. Relative to U.S. firms, however, there is much greater emphasis on improving work practices. This is reflective of the behavioral approach to promoting safety and health in Japan as compared with the engineering approach in the United States.

Training. Jidosha conducts a variety of training programs for its employees. Including on-the-job training, during which safety and health are strongly emphasized, training for new employees lasts one year. Describing Jidosha's overall philosophy of training, a Jidosha USA supervisor noted that "once you've trained somebody in the job . . . if they make a mistake, it's not their fault, it's the trainer's fault. Because you did not train them enough, you did not explain to them enough, you did not take the time out to listen to them. That in itself creates a team concept because everything is shared." Managers estimate that about 10 percent of the total training time is devoted to training related to safety and health.

In addition to training its employees, Jidosha, encouraged by the Japanese government, trains its subcontractors and suppliers in safety and health procedures. Representatives of these firms receive training from two full-time safety and health instructors in the Purchasing Division of the main plant who receive assistance from corporate- and plant-level safety and health personnel.

Committees. Safety and health committees function at virtually every level of the company and include representatives of the various units that have responsibilities for safety and health. At the corporate level, there is a General Safety and Health Committee, which is chaired by the director of industrial relations. It is composed of plant-level and corporate-level managers with line responsibility for safety and health, as well as union officials, who comprise approximately half the committee. The committee meets once a year to review results of the preceding year and to set broad safety and health goals for the coming year.

Plant-level safety and health committees meet monthly at the main plants and are chaired by the plant manager. They are composed of division managers, general managers, and union representatives, who account for approximately half the membership. Like the corporate-level committee, they are responsible for goal setting and performance evaluation regarding safety and health. Because they meet more frequently than the corporate-level committee, their role in the management and coordination of safety and health activities can be more active.

Division-level committees are similar to plant committees except that the union is less represented. Out of fifteen members, only one to three are union representatives. Division committees meet every other month.

Each department manager chairs a safety and health council, which meets monthly. These councils include foremen and plant safety and health officials (superintendents, Work Safety Section) and one to three union representatives. Section meetings consisting of section superintendents, foremen, and some employees are also held monthly.

Every work team holds a safety and health meeting in the morning on the third Wednesday of the month. In addition, many divisions have voluntary safety and health meetings in the morning before work and, in some cases, after work (focusing on safe commuting). Morning meetings last just a few minutes and are led by work-team leaders appointed by foremen. In some workshops, workers set goals that go beyond avoiding accidents, such as meeting a specific schedule of training programs. Progress toward these goals is evaluated periodically at workshop meetings.

Safety and health staff serve on safety and health committees at various levels of the organization. They facilitate the activities of these committees by providing information about companywide safety and health issues.

Signs. Signs are used extensively to promote proper work practices, although safety and health officials concede that it is difficult to measure their effectiveness. Most of the signs are similar to those used in U.S.-owned plants. One exception is that divisions, departments, and workshops of Jidosha display tallies showing the number of workdays without a lost-work injury. Tallies showing the number of days since the last accident are commonly used in the United States but usually at the plant level only. Jidosha's safety records are admirable, and these tallies are very impressive. The sign in the Body Division, for example, indicated that there had not been an accident on either the day or night shift since it began production more than seventeen hundred days (almost five years) earlier.

Inspections. Reflecting the philosophy that safety, quality, and productivity are mutually consistent goals, workers in Japanese companies are expected to inspect for safety hazards in their work areas in much the same way as they inspect for product defects. They are instructed to correct the problem immediately if they can or to call their supervisor using their yellow or red signal light. Inspecting for safety hazards is part of the workers' general housekeeping and maintenance responsibilities. Evidence suggests that workers take this responsibility seriously. Work areas are very clean and generally free of hazards.

Formal inspections are also conducted on a regular basis by plant, division, and corporate safety and health staffs and by committees at various levels of the organization. In some divisions, inspections are conducted daily.

From time to time, union representatives also undertake formal inspections. These inspections are in addition to the day-to-day monitoring union representatives conduct on the production line.

Hazard-Prediction Activities. Hazard-prediction activities or hazard-prevention activities (abbreviated KYT for the Japanese-English phrase *Kiken Yochi* Training) refer to a broad range of activities workers are trained to undertake to avoid hazardous situations. These activities are promoted by JISHA and are described in detail by Hajime Tanabe (1984).

Hazard-prediction activities have been given great emphasis at Jidosha in recent years, reflecting the behavioral orientation of the safety and health system. The most obvious of these activities is the safety check or "finger point and call." After the voluntary morning exercise, workers stand opposite one another and, in a procedure that lasts only a few seconds, they point first to their face (checking to see that the worker opposite them has a healthy color) and then to their chest, waist, legs, and feet, each time calling out "*yoshi*" (meaning okay! alright! or let's go!). This exercise is essentially a ritual. When asked what one says to one's counterpart if he or she is not alright, safety and health staffers said that the worker would still say "*yoshi*" and then take him or her aside to discuss the problem.

The safety check has received increased emphasis in recent years as a way to detect sicknesses or injuries workers might otherwise try to hide. Like the morning exercise, the check is recommended but not required. Unlike the morning exercise, however, workers did not participate in large numbers. An estimated 70 to 80 percent of production and office workers combined participated in the morning exercise, but far fewer participated in the safety check. Among office workers, only safety and health staff members participated, and they may have been doing it for my benefit.

Jidosha workers are also trained to use the finger point and call procedure when undertaking potentially hazardous activities such as crossing a street. I did not observe employees other than safety and health staff using the technique, however. In fact, some managers, and even the union president, indicated that workers felt embarrassed or self-conscious about using it. By contrast, train engineers on the public rail system invariably use the technique before proceeding through intersections.

Among other hazard-prediction activities are SKY (immediate hazard prediction) activities for use in nonstandardized work situations.[8] SKY activities are being promoted by JISHA and have been adopted by Jidosha. An example of this new technique is the procedure to be followed when a worker enters a machinery area to perform maintenance and repair work. Jidosha has developed a SKY card that reads "Danger, don't touch any switch, operator is in the machine." The Jidosha worker shuts down the machine, takes out the key, and puts this sign by the control panel when he or she enters the area.

The SKY card also instructs the worker to perform the following four-step mental exercise when entering a machine area. First, he or she should mentally prepare a list of three dangerous situations that could be encountered while doing the maintenance. Second, he or she should determine which of these three situations is the most dangerous. Third, he or she should think of three safety measures to counteract the most dangerous of these situations. Fourth, he or she should determine which is the best safety measure and verbalize it out loud in an action statement in the following form: "When ______, let's ______, ok."

Workers at U.S. companies usually shut down a machine and put a lock on the control switch so that it cannot be started. (In more hazardous situations, Jidosha workers use a similar interlock system.) The problem with the U.S. system is that workers do not always follow the procedure, either because they do not have the lock and key with them or because they think they will be in the machine area for only a few seconds. Injuries occur then when another worker starts up the machine without knowing the first worker is in it.

Jidosha USA safety and health staffers were impressed with Jidosha's system and thought it was sufficient in most situations. While being trained in this procedure in Japan, one Jidosha USA safety and health specialist commented that "other workers have keys that would fit [the machine], but that's murder." Nevertheless, as I discuss in the following chapter, this comment reflected an overly optimistic view of the ability or willingness of U.S. workers to follow work safety rules.

Injury and Illness Reports and Investigations. Injury and illness statistics are collected from all of Jidosha's operations in Japan. They are assembled and disseminated monthly to line managers from foremen on up. The foremen then distribute them among workers. Staff managers at the level of assistant department head and up receive injury and illness reports monthly, and are instructed to share them with subordinates.

These managers are also given reports showing injury and illness experiences for the current and previous years.

Several reporting systems are used, creating confusion about the criteria for reporting. Lost-work injuries and illnesses are reported to the government in one of three categories: those involving absence from work for a portion of a day up to three days; those entailing four or more days' absence from work; and those involving more than eight days' absence from work. For workers' compensation purposes, the company also reports any injuries or illnesses requiring clinical treatment. Those not involving any absence are counted as nonleave injuries and illnesses. Thus, a worker who is unable to finish the day's work because of an injury or illness requiring clinical treatment would be counted as having a lost-work injury or illness under the workers' compensation system even if he or she returned to work the following day. Injuries requiring only first-aid treatment on the shop floor are counted as first-aid injuries and are not reported to the government. Further complicating the issue, another government reporting system uses one or more days off (not including the day of the accident) to define a lost-work injury.

Perhaps because of the number of categories and criteria, line managers seemed confused regarding what constituted a reportable work injury or illness. In one subcontractor firm, managers indicated that an injury preventing a worker from finishing the workday would not be counted as an injury if the worker returned to work the following day. Actually, it should have been counted as a nonleave injury. Jidosha managers expressed different views about whether an injury that requires a worker to stay at the hospital for treatment but to return to work the following day should be counted as a nonleave injury or as a lost-work injury. Answers to many such questions at Jidosha and elsewhere were not always consistent even among safety and health staff.

Jidosha also collects injury and illness statistics from its principal subcontractors and from its subsidiaries. Ninety-two of these subcontractors and subsidiaries with employment ranging from 9 to 1,397 comprise an organization I shall call the Jidosha Friendship Association (JFA), which Jidosha helped organize and continues to support. Together, the firms in JFA have more than twenty-three thousand employees (close to the number at Jidosha itself). JFA in cooperation with Jidosha collects and disseminates injury and illness statistics monthly and yearly and gives out safety and health awards at an annual banquet. Some members also meet monthly to share information on safety and health activities.

The Labor Standards Bureau encourages master employers to collect injury and illness statistics from its subcontractors. Like many such requests in Japan, the government relies on voluntary compliance by employers rather than the threat of penalties.

Below the first layer of primary subcontractors who belong to JFA, there are a great many secondary and tertiary, typically smaller subcontractors. Jidosha purposely avoids contact with them regarding safety and health matters, at least in part out of fear of being held responsible for improving conditions in their facilities. When I suggested collecting information on their safety and health activities and injury and illness statistics, the general supervisor of safety and health said he did not want to know about these matters. He conceded that officials at the prefectural Labor Standards Bureau would be "very happy" if he collected such information, but he was concerned that they would then hold Jidosha responsible for making improvements. According to him, "Jidosha can't afford this." (Theoretically at least, first-level subcontractors oversee the safety and health activities of second-level subcontractors, second-level subcontractors oversee the activities of the third level, and so on.) The avoidance by Jidosha personnel of direct contact with lower-level subcontractors reflects the tendency common in Japanese corporations and society at large to make sharp distinctions between one's responsibilities to those in one's group (however that group is defined) and to those outside it.

Master employer–subcontractor relationships relative to safety and health vary among firms in the Japanese automotive industry. A distinction is sometimes made between subcontractors in which the master employer has an equity position and those in which it does not. Representatives of another automotive firm contended that they would not be permitted to collect injury and illness statistics from subcontractors in which they had no ownership. Jidosha officials cited no such prohibition, however; Jidosha routinely collects injury and illness data from JFA firms that are not subsidiaries. No doubt the family atmosphere promoted by JFA facilitates this sharing of information.

When a serious accident occurs at Jidosha, individuals at several levels of the organization compile comprehensive reports. These include corporate safety and health staff, as well as division safety and health staff and individual line managers. A specialist from the Jidosha hospital investigates the medical aspects of the injuries, and teams of investigators from the area in which the accident took place, from the Safety and Health Department, and from the Medical Department review all aspects of the accident. Individuals with knowledge of the incident are interviewed, pictures are taken of the accident site, and ultimately a thick dossier is assembled. Condensed reports are issued throughout the company, including to overseas locations, if it is determined that changes in production machinery or processes are necessary.

Other Safety and Health Activities. In addition to the primary safety and health activities reported above, Jidosha undertakes several ancillary activities. In another symbolic exercise, safety and health awards are made annually to work groups based on their safety and health performance. These awards are for only token amounts, 200 yen (approximately $1.35) per worker, but the honor of winning apparently motivates work groups to improve their performance. The award also reflects favorably on the supervisor of the winning group. The company's safety and health awards for work groups are based on zero-accident experience as well as ten safety and health performance criteria. Awards are given for excellent results, superior results, and good effort. The number of groups receiving awards in these three categories were two, six, and six respectively in the 1985–86 fiscal year.

The Health Care Center Division conducts four categories of activities that support the activities discussed above: (1) medical examinations, (2) statistical monitoring of diseases, (3) monitoring of workshops for hazards, and (4) disease control. Health examinations are provided (as required by law) at recruitment, at employment, and at yearly intervals. Workers assigned to hazardous jobs receive examinations more frequently and receive special examinations at the ages of thirty-five, forty, and yearly thereafter. These are used to check not only for evidence of diseases but also overall physical fitness. The statistical monitoring of diseases provides a method whereby the company can determine the existence of hazards in certain areas or jobs. This process is done in addition to the inspections of workshops conducted by industrial physicians and others in the division.

The last category, disease control, refers not only to the elimination or reduction of hazards but the promotion of healthful lifestyles. Workers are encouraged to exercise, eat a proper diet, get adequate sleep, and moderate harmful activities such as smoking and drinking. Through an off-the-job program called the "Get a Hundred Campaign," points are assigned for performing prescribed exercises on one's own time. For example, one receives three points for jogging for fifteen minutes or more, swimming for twenty minutes or more, or playing a round of golf. One can earn a maximum of five points in a day. Thus, workers are required to exercise approximately two out of three days to make the target of one hundred points in a month.

In contrast to this rather rigorous exercise schedule, the anti-smoking activities at Jidosha seem halfhearted. Given the high levels of cigarette consumption in Japan[9] and the apparent belief that trying to get workers to stop smoking altogether would be futile, Jidosha, like other auto companies, sponsors smoking-reduction campaigns. Jidosha's relatively weak efforts in this area are consistent with the timid warning on packages of Japanese cigarettes: "For your health, don't smoke too much." Vigorous anti-smoking campaigns such as those in some U.S. companies are very rare in Japan. Breaks are referred to as *kitsuen jikan* or "smoking time" at Jidosha and at other companies. Company officials concede that the effectiveness of the smoking-reduction campaign has been "slow in coming."

The consumption of large amounts of alcohol after work with one's colleagues is a time-honored tradition in Japan. It is viewed as a socially acceptable release from pressure and the strictly prescribed norms of behavior in Japanese society. It is said to enhance camaraderie among work groups and to improve communication among workers and between workers and their superiors. According to Japanese custom, one is not held accountable for minor indiscretions committed while intoxicated. Workers who are drinking with their boss are thus able to complain about undesirable aspects of work they would never dare discuss in any other context. Aggressive driving (some would say bad driving) is another accepted release from prescribed norms of behavior. Recognizing that it can be very dangerous, especially when combined with drinking, Japan tightened its drunk driving laws and penalties in the mid–1980s. This has had a moderating effect on public drinking, especially among workers who cannot use public transportation, as is the case for many Jidosha employees. Increased use of car pooling with a designated nondrinking driver and an overnight stay in a moderately priced business hotel are two favored responses to the laws that enable workers to continue the tradition of drinking after work.

In light of the increasing use of drug and alcohol tests for prospective employees and random testing of employees by U.S. companies, it is noteworthy that, in interviews, Jidosha's managers summarily rejected using such practices in Jidosha's Japanese operations. (Preemployment drug and alcohol tests are, in fact, used at Jidosha USA.) Jidosha officials contend that drug or alcohol abuse is not a problem among their work force and that only one worker was ever found drinking on the job.[10] They also contend that of the twenty-five thousand or so workers, only fifteen are alcoholics, of whom seven require medical care. Certainly the much closer confines of the work environment and the more intense supervision and closer observation of co-workers minimizes the need for drug and alcohol tests. It would be difficult to find a private place at work to take a drink. Moreover, managers feel that the tests would disrupt the harmony of labor-management relations. In keeping with the within-group/out-of-group distinction, preemployment drug and alcohol testing would probably be considered acceptable if drug and alcohol abuse were considered a more serious problem.

Denial of the existence of an alcohol problem, or of any personal problems, is a fairly typical Japanese response. The Japanese downplay the excessive use of alcohol *off the job*, which is evident to anyone who has been in a Japanese bar (or walked by one at night); they contend that American reporters exaggerate the extent of the problem. Another assertion by Jidosha officials exemplifies this tendency to deny the problem: "Alcohol abuse is inconceivable among workers. The alcoholic rate is much higher among Americans and alcohol abuse, if any, is rather on the American side. No grave concern about this matter exists among Jidosha employees."[11]

The high level of alcohol abuse, as well as Japan's suicide rates, suggest that many workers have serious personal problems. Whether Japanese workers would accept an employee assistance program is another matter. Psychological counseling is not nearly as commonplace in Japan as it is in the United States, and it carries more of a stigma. According to a leading Japanese psychiatrist who specializes in work-related disorders, the quality of mental health care available in Japan is very uneven. Instead of referring an employee to a professional, the standard practice is to hold the manager who hired the employee responsible for looking after him or her (Kaneko 1986). Satoshi Kamata (1986) reported that Toyota has decreased the number of stress-related suicides among its workers by following such a practice. Greater attention has been paid to mental health issues, however, since a Japan Air Lines pilot suffered a nervous breakdown and crashed his plane into Tokyo Bay (Kaneko 1986).

The Japanese put a great deal of emphasis on preventing illnesses and health conditions brought on by advancing age, such as high blood pressure, obesity, diabetes, and liver disease. This is especially important because the average age of the workers at Jidosha and in the rest of the industry is increasing very rapidly. The average age at Jidosha was 39.2 years in 1985, compared with 35 years in 1976. This rate of increase exceeds the rate of aging of Japanese society as a whole.

In sum, Jidosha's corporate safety and health staff think it is important to emphasize "human aspects of accident prevention." This emphasis is reflected in each of the above categories of activities and in the union activities discussed below. The safety and health staff at Jidosha believe that the physical conditions in the workplace are already good and that only marginal improvements will be made in the future. In general, workers in Jidosha are more willing to cooperate with this behavioral approach.

Union Activities. Like other Japanese unions, the Jidosha Workers' Union identifies its interests as similar to those of Jidosha, although its agenda in the area of safety and health is moderately independent. According to a publication entitled *Union Activities at Jidosha*:

> The Jidosha Workers' Union flexibly copes with production needs based on the position that the fundamental stability of the company leads to stable employment and stable lives for union members. However, it is important to clearly establish the limits of cooperation to protect the health of workers. Therefore, the Jidosha Workers' Union aims at a harmonious balance between the economic rationality of the company and the health and quality of life of the union members (Jidosha Workers' Union 1986:14).

Consistent with this philosophy, the union undertakes essentially interdependent activities in the occupational safety and health arena. For example, union safety and health representatives at the workshop level look for hazards and can reprimand workers who engage in unsafe acts. But, unlike the situation in U.S. companies, where there are usually one or two full-time safety and health representatives, union stewards at Jidosha are responsible for safety and health part time. They also represent workers in other union matters and have full-time production jobs. There are approximately 140 representatives in the main Jidosha plant, which, compared with U.S. companies, gives the union more coverage of the work environment. The representatives are not as well trained in safety and health, however, and do not have as much time to devote to these activities as UAW safety and health representatives.

In addition to the union representatives at the workshop level, one full-time union member is a safety and health expert and other union representatives who do not have factory jobs have some safety and health responsibilities. The full-time expert receives periodic training in government safety and health rules, and other union safety and health representatives sometimes do as well. (In any event, the full-time expert would pass along this information.)

The union awards its own safety and health prizes, also for 200 yen. The award is based on two criteria: accident experience and safety and health–promotion activities. To win, members of a work group must have had no accidents during the year, either at work or while commuting, and the workshop's safety and health activities must be judged to be superior. Work groups can win either union or company safety and health awards or both.

The union also conducts safety patrols (inspections) monthly and holds a Jidosha World Auto Council meeting every other year, where representatives from several Jidosha locations in Asia (Thailand, Malaysia, Taiwan, Korea, and Japan) get together to discuss issues such as job security as well as safety and health issues. The union sponsors other safety- and health-related activities, including workshop safety and health–promotion committees and contests focusing on safety at home.

As is the practice in Japan, safety and health per se are not subjects of labor-management negotiations at Jidosha. Negotiations pertaining to a potentially contentious issue having safety and health implications are handled by a special-duty negotiation team. The team also meets whenever monthly production schedules call for overtime work in excess of limits set in the contract. The union specifically cites overtime as an area in which the economic concerns of the company and the health concerns of the workers need to be kept in balance (Jidosha Workers' Union 1986). There is little room for negotiation, however. The maximum hours of overtime allowed by the contract are 2 hours per day in the conveyor shop and 2.25 hours per day in other shops of the manufacturing division, and those set by labor laws are 2.5 hours per day for manufacturing operations. Within this narrow range, the union views "excessive" overtime as beyond the limits of cooperation.

Union and company officials generally agreed that their safety and health programs improved markedly during the 1980s. Before then, the activities were less intense and there was less emphasis on training. Further, the safety and health function was not as well staffed. Improvement in these areas was motivated by a very poor safety and health record in the 1978–79 fiscal year. There were three fatalities that year, and union leadership and management agreed that greater effort was needed to improve safety and health conditions.

Injury and Illness Experience

Long-term trends in injury and illness experience for Jidosha and for the entire Japanese auto industry (the ten largest master employers, excluding Jidosha) are shown in figures 4-1 and 4-2 and in tables 4-1 and 4-2. Statistics are provided for fiscal year 1967–68 to 1987–88 for Jidosha and from calendar year 1968 to 1988 for the rest of the auto industry.[12] These statistics include the lost-work injury and illness (leave accident or work shutdown) frequency rate (number of lost-work injuries and illnesses per million hours worked) and the intensification or severity rate (number of labor days lost due to lost-work injuries per thousand labor hours, with a work fatality charged as seventy-five hundred lost-work days in the year of occurrence; permanent disabilities are also charged as lost-work days, with the number determined by the severity of the disability).

Industry statistics are collected and disseminated to members by a safety and health subcommittee of the Japanese Automobile Manufacturers' Association. The safety and health managers of firms represented on this subcommittee meet regularly to discuss matters of mutual concern. The availability of the injury and illness statistics that they share encourages competition among them to reduce injury and illness rates.

As table 4-1 indicates, Jidosha's reported injury and illness rates have declined significantly since 1968. The total frequency rate started the 1968–88 period with a high of 16.36 cases per million hours worked. There were year-to-year declines in sixteen of the following twenty-one fiscal years, and it hit a historic low in 1988, the most recent year for which data are available, of .79 injuries and illnesses per million hours worked. This is 5 percent of the 1968 level, and it means that there was less than one injury or illness requiring medical attention for every six hundred full-time workers.

Table 4-1 also shows an overall decline in the severity rate, although it has been highly variable. Jidosha started the period with a severity rate of .38 and ended the period with a severity rate of .159. The peaks of .45, .57, and .48 in 1971, 1979, and 1981, respectively, can be attributed largely to work fatalities in these years, but the long-term decline still does not compare favorably with declines in other injury and illness rates at Jidosha. The severity rate in 1988 was 42 percent of the level in 1968 but as low as 5 percent in 1986.

As shown in table 4-1 and figure 4-1, there have also been impressive declines in the work shutdown or lost-work injury and illness frequency rate, although most of the decline took place before 1982. The company started the 1968–88 period with a high of 1.85 lost-work injuries and illnesses per million hours worked and ended the period with a rate of .31. The long-term declines in this rate are not as impressive as they are for the total injury and illness frequency rate. (The 1988 rate is 17 percent of the 1968 high.) Nonetheless, on average, there was less than one lost-work injury and illness for approximately sixteen hundred or so full-time workers.

Jidosha injury and illness rates are not quite so impressive when compared with those of other Japanese auto companies. As can be seen in figure 4-1 and table 4-2, Jidosha's lost-work injury and illness frequency rate was higher than the average of the other auto companies in every year except 1968 and 1974. In many years, Jidosha's rates were only slightly higher, but in 1978, 1979, and 1980, they were 90 percent,

Table 4-1. Jidosha Injury and Illness Experience

Year[a]	Total Labor Hours (10,000s)	Deaths	Amputations	Total Number of Injuries and Illnesses	Number of Lost-Work Injuries and Illnesses	Number of Lost Days[b]	Total Injury and Illness Frequency Rate[c]	Lost-Work Injury and Illness Frequency Rate[d]	Intensificatio (Severity) Rate[e]
1968	6,375	1	41	1,043	118	24,442	16.36	1.85	0.38
1969	5,929	0	19	804	101	9,303	13.56	1.70	0.16
1970	6,013	0	20	741	92	7,630	12.33	1.53	0.13
1971	6,385	3	17	729	91	28,433	11.42	1.43	0.45
1972	7,410	0	26	842	122	9,882	11.36	1.65	0.13
1973	8,057	0	22	766	117	9,029	9.51	1.45	0.11
1974	7,131	0	15	401	73	10,272	5.62	1.02	0.14
1975	6,371	1	13	290	78	11,034	4.55	1.22	0.17
1976	6,372	0	8	257	58	2,601	4.03	0.91	0.04
1977	6,044	0	4	283	68	3,895	4.68	1.12	0.06
1978	4,999	0	13	214	79	8,825	4.28	1.58	0.18
1979	4,682	3	14	210	67	26,716	4.48	1.43	0.57
1980	5,179	0	21	183	66	6,976	3.53	1.27	0.13
1981	5,537	2	16	163	66	26,544	2.94	1.19	0.48
1982	5,559	0	10	111	27	7,037	2.00	0.49	0.13
1983	5,559	0	11	88	33	10,535	1.58	0.59	0.19
1984	5,745	1	2	48	23	13,663	0.84	0.40	0.24
1985	6,199	0	9	63	29	1,963	1.02	0.47	0.03
1986	6,362	0	3	61	29	1,346	0.96	0.46	0.02
1987	5,902	1	6	58	18	10,235	0.98	0.30	0.17
1988	5,849	1	4	46	18	9,317	0.79	0.31	0.159

Source: Jidosha company records.

[a]Data for 1968 to 1986 are for fiscal years; it is uncertain whether data for 1987 and 1988 are for fiscal years or calendar years.

[b]Each fatality is charged as 7,500 lost days; each amputation is also charged as lost days, with the number depending on the degree of severity.

[c]Number of injuries and illnesses per million hours worked.

[d]Number of leave injuries and illnesses per million hours worked.

[e]Number of lost-labor days due to injuries and illnesses per thousand hours worked.

Figure 4-1. Jidosha and Industry Leave Accident Frequency (Work Shutdown) Rate

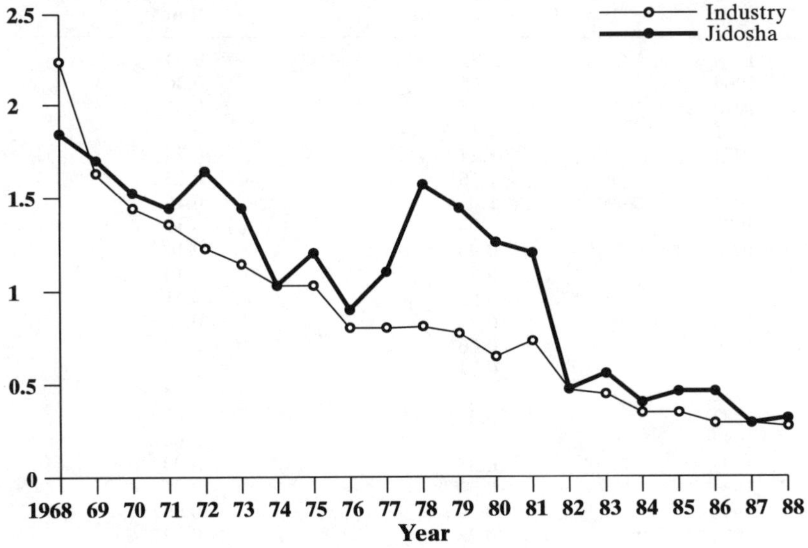

Source: Jidosha and Japanese Automobile Manufacturers' Association.

Leave accident rate = Leave (lost-time) injuries and illnesses per million hours worked.

86 percent, and 92 percent higher, respectively. By 1988, the discrepancy had narrowed to 11 percent, however, and Jidosha ranked sixth of ten auto firms on this rate.

Comparisons with other auto companies are somewhat more favorable when severity rates are considered (fig. 4-2 and table 4-2). Jidosha's severity rate was lower than that of the other auto companies in eleven of twenty-one years considered. It exceeded that of the other companies in nine years, and in one year the rates were the same. In the most recent year, Jidosha's severity rate, .159, was more than twice that of the industry, ranking it eighth of the ten auto companies. Relatively poor performance in 1987 and 1988 can be attributed in large part to a fatality in each of these years.

Because the occurrence or nonoccurrence of serious or fatal injuries can cause large year-to-year fluctuations in the severity rate, it is perhaps more enlightening to consider this rate over several years. The average for Jidosha over the last four years for which data are available, 1985–

Table 4-2. Injury and Illness Experiences of the Japanese Auto Industry[a]

	Lost-Work Injury and Illness Frequency Rate[b]	Intensification (Severity) Rate[c]
1968	2.23	0.54
1969	1.65	0.36
1970	1.45	0.31
1971	1.37	0.31
1972	1.24	0.25
1973	1.16	0.25
1974	1.03	0.16
1975	1.04	0.17
1976	0.90	0.11
1977	0.81	0.13
1978	0.83	0.11
1979	0.77	0.16
1980	0.66	0.12
1981	0.74	0.13
1982	0.47	0.14
1983	0.44	0.09
1984	0.34	0.04
1985	0.35	0.09
1986	0.30	0.04
1987	0.29	0.067
1988	0.28	0.077

Source: Japanese Automobile Manufacturers' Association.
[a]Figures are the averages for the ten largest Japanese automotive firms, excluding Jidosha.
[b]Number of lost-work injuries and illnesses per million hours worked.
[c]Number of lost-labor days per thousand hours worked.

88, was .09475, whereas the average for the other nine Japanese auto companies was .0685. Thus, Jidosha's average was still 38 percent higher than the average for the other firms.

One should be cautious, however, not to place too much emphasis on the above intraindustry comparisons. Differences exist among the auto companies in injury and illness reporting and recording policies and practices, including the definitions of work injury and illness, and some practices, quite frankly, are strange. For example, as mentioned in the previous chapter, one firm records in a total-injury category only those injuries not involving days away from work which take more than six days to heal. An injury that heals within six days is not considered a nonleave injury.

Another less bizarre but more significant irregularity arises because companies do not record all the injuries and illnesses that occur. Although all work-related injuries and illnesses receiving medical treatment are supposed to be reported to the government for workers' compensation purposes, auto industry safety and health officials concede that large companies that have their own hospitals do not report all the

Figure 4-2. Jidosha and Industry Severity (Intensification) Rate

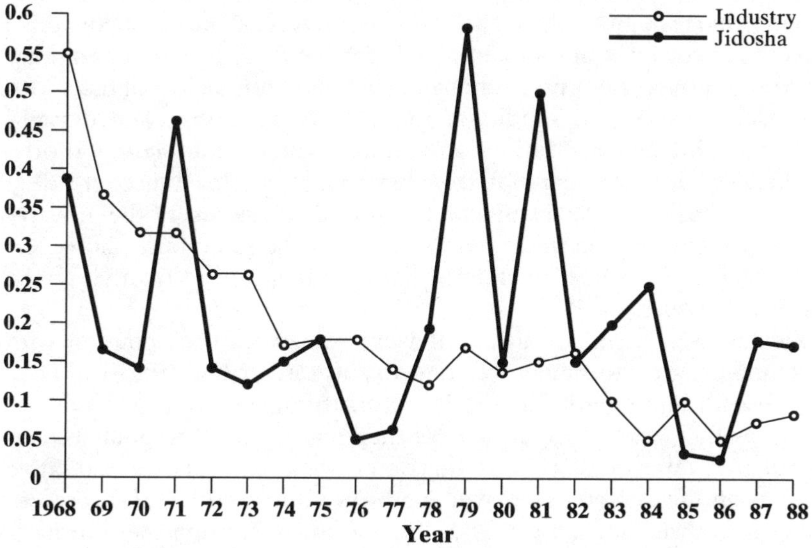

Source: Jidosha and Japanese Automobile Manufacturers' Association.

Severity rate = Number of lost-labor days per thousand labor hours for the ten largest Japanese automotive firms, excluding Jidosha.

work-related injuries and illnesses they treat. These firms thus appear to have safer and more healthful conditions than they do.

Another practice that makes comparisons difficult is that firms often send workers to company hospitals for rehabilitation during working hours but do not consider them absent. Thus, their absence is not reflected in the severity rate, and their condition is not recorded as a lost-work injury or illness.[13] Although there is legislative and regulatory authority to levy fines for misrecording injuries and illnesses, large government penalties such as those imposed by OSHA in recent years are unheard of in Japan. Rarely if ever are any fines imposed for misrecording safety and health statistics.

Further distortions in industry injury and illness rates occur because workers conceal injuries and illnesses. Some workers simply attempt to conceal the injury or illness; others claim that the condition is not work-related (e.g., that a work injury occurred at home). The magnitude of the underreporting is impossible to determine, but safety and health managers throughout the industry are suspicious of statistics reported by their competitors. One's relative standing on injury and illness performance, which is known throughout the industry, is a matter of collective concern for corporations and especially for their safety and health managers. A relatively poor performance entails a corporate loss of face. Mutual suspicion about the validity of others' statistics probably encourages further fudging by corporations attempting to match their competitors.

Jidosha's statistics bear out the view that its performance in safety and health has improved significantly over time. The rest of the industry has been improving its performance at about the same pace, however, so that Jidosha has been unable to improve its position relative to the rest of the industry.

Jidosha's safety and health staff were somewhat disappointed with the company's performance relative to the rest of the industry. They were also disappointed that the lost-work frequency rate was not declining as they had expected and as they had targeted in goal setting. Of much greater concern was the deterioration of Jidosha's relative position on the severity rate over time. As safety and health representatives acknowledged, injury and illness rates at Jidosha seem to have hit a plateau—recent declines are relatively minor compared to those that occurred earlier. The two most recent fatalities occurred after I left Japan, but they are indicative of problems in the safety and health system and undoubtedly were cause for major concern in Jidosha.

In addition to injury and illness statistics, Jidosha keeps comprehensive statistics on the health of its workers. Two measures are used. One measures the worker's physical age (based on physical condition), and the other groups workers aged thirty-five and forty or older into one of four categories of healthfulness: very healthy, "slight abnormality" but basically healthy, questionable (merits observation), and ill (must receive clinical treatment). Classification in one of the categories other than "very healthy" may be for work-related or nonwork-related reasons.

Jidosha has made significant progress in improving workers' performance on the first of these measures. In 1980, workers, on average, had a physical age 5 years older than their chronological age. By 1986, however, their physical age was 1.2 years younger than their chronological age. Thus, a typical worker who had a chronological age of 35 in 1980 and a physical age of 40 would have had a chronological age of 41 by 1986 but a physical age of only 39.8.

I took the fitness test to determine my own physical age. The test consisted of a standing high jump, an agility drill, a grip-strength test, and a cardiovascular recovery test. Although I engage in a wide variety of sports and am a competitive distance runner, my physical age and my chronological age were the same, 36. Apparently, Jidosha workers are in exceptionally good condition, at least those who took the test. Some managers who were overweight and obviously not in good shape conceded that they avoided taking it. One public relations manager who had avoided taking it blamed his poor condition on the amount of entertaining he is required to do at bars and restaurants as part of his job.

There has also been improvement in the second of the two health measures. The percentage of workers in the "very healthy" category increased from 20.8 percent to 31.3 percent from 1984 to 1986, while the percentage in the "ill" category decreased from 18.0 percent to 16.5 percent. New workers to the "very healthy" category appear to have come from the "slight abnormality" group, which declined from 30.5 percent in 1984 to 9.7 percent in 1986.

The improvements in both of these measures were attributed to modifications in the work environment, more frequent check-ups, and the increased emphasis on the importance of proper diet, exercise, and sleep and the avoidance of unhealthful habits such as excessive smoking and drinking.

Injury and Illness Experience of the Subcontractors

Tables 4-3 and 4-4, provided by Jidosha, list the injury and illness statistics for the 1985 calendar year for each of the ninety-two principal subcontractors that belong to the Jidosha Friendship Association. These subcontractors are grouped into three categories: stamping, machining, and components. Aggregate statistics are provided for these three groups for 1982–85, as well as statistics on firm size.

There is a great degree of variation in size among JFA firms. One has only 9 workers, and two have more than 1,300 workers each. The average in 1985 was 239 workers.

On average, the injury and illness rates for the subcontractors were substantially higher than those for Jidosha. The 1985 rate for the subcontractors was 1.91 lost-work injuries and illnesses per million hours worked, compared to .46 for Jidosha in the same year.

The severity rates, relatively speaking, were even worse. In calendar year 1985, the subcontractors had a rate of .22 lost workdays due to injuries and illnesses per thousand labor hours, more than seven times Jidosha's rate. Performance on this dimension was not significantly worse, however, than Jidosha's in previous years. The subcontractors' severity rate in 1985 was approximately the same as Jidosha's rate in 1983–84.

There was considerable variation among the subcontractors in their injury and illness rates. Frequency rates ranged from a low of zero lost-

Table 4-3. Injuries and Illnesses at Jidosha Subcontractors, 1985

Company	Number of Workers	Number of Cases	Lost Workdays	Number of Hours Worked	Frequency[a] Rate	Severity[b] Rate	Nonfatal Cases without Lost Workdays
Category I: Machining Subcontractors							
1	22	0	0	41,952	0	0	0
2	28	1	8	66,951	14.94	0.12	0
3	31	3	17	64,256	46.69	0.26	0
4	40	0	0	85,683	0	0	0
5	166	1	49	381,053	2.62	0.13	4
6	165	2	1,066	393,986	5.08	2.71	1
7	113	1	11	263,583	3.79	0.04	1
8	640	3	135	1,516,342	1.98	0.09	38
9	93	0	0	220,125	0	0	0
10	860	2	425	2,177,148	0.92	0.20	5
11	676	0	0	1,552,765	0	0	29
12	148	1	50	369,619	2.71	0.14	0
13	30	0	0	70,902	0	0	0
14	114	0	0	288,372	0	0	12
15	415	0	0	963,448	0	0	23
16	23	0	0	58,665	0	0	0
17	303	0	0	630,426	0	0	3
18	121	2	14	236,771	8.45	0.06	0
19	72	0	0	168,190	0	0	1
20	36	1	8	82,291	12.15	0.10	1
21	53	0	0	141,240	0	0	0
22	54	0	0	118,346	0	0	1
23	149	1	50	345,546	2.89	0.14	0
24	25	1	52	51,873	19.28	1.00	5
25	55	3	649	137,893	21.76	4.71	1
26	24	0	0	53,065	0	0	0
27	323	0	0	804,888	0	0	10

28	20	0	0	48,204	0	0	0
29	45	1	600	99,360	10.06	6.04	0
30	307	1	17	730,666	1.37	0.02	13
31	75	1	34	175,747	5.69	0.19	0
32	34	0	0	88,770	0	0	0
33	15	0	0	37,296	0	0	0
34	28	0	0	66,820	0	0	0
35	85	1	112	198,303	5.04	0.56	7
36	31	0	0	75,902	0	0	2
37	113	0	0	297,234	0	0	1
38	172	1	6	374,476	2.67	0.02	7
39	790	0	0	1,893,199	0	0	2
40	489	5	846	1,096,657	4.56	0.77	13
Subtotal	6,983	32	4,149	16,468,012			180
Average	174.6	.8	103.7	411,700	1.94	0.25	4.5
Category II: Stamping Subcontractors							
1	42	0	0	97,109	0	0	0
2	151	1	7	290,043	3.45	0.02	9
3	734	3	1,127	1,624,933	1.85	0.69	1
4	163	1	50	409,490	2.44	0.12	22
5	566	4	262	1,245,218	3.21	0.21	9
6	15	0	0	28,427	0	0	0
7	66	1	6	162,560	6.15	3.61	4
8	99	2	85	234,598	8.53	0.36	1
9	159	4	73	343,794	11.63	0.21	17
10	88	0	0	204,292	0	0	3
11	14	1	20	34,241	29.20	0.58	0
12	194	4	707	482,282	8.29	1.47	13
13	145	0	0	352,495	0	0	0
14	80	0	0	191,382	0	0	0
15	33	1	400	72,954	13.71	5.48	0
16	142	1	50	312,232	3.20	0.16	1
17	41	3	142	110,040	27.26	1.29	0

Table 4-3. ***Continued***

Company	Number of Workers	Number of Cases	Lost Workdays	Number of Hours Worked	Frequency[a] Rate	Severity[b] Rate	Nonfatal Cases without Lost Workdays
18	37	0	0	83,001	0	0	0
19	927	1	67	2,214,323	0.45	0.03	20
20	100	1	100	193,505	5.17	0.52	0
21	221	3	93	483,149	6.21	0.19	4
22	55	2	1,000	106,798	18.73	9.36	1
23	301	0	0	657,368	0	0	8
24	407	2	36	944,405	2.12	0.04	7
25	81	1	51	199,512	5.01	0.26	0
26	660	3	713	1,533,450	1.96	0.46	6
27	552	1	5	1,264,702	0.79	0.004	10
Subtotal	6,073	40	4,994	13,876,303			136
Average	224.9	1.48	185.0	513,937	2.88	0.36	5.04
Category III: Component Manufacturing Subcontractors							
1	206	0	0	463,489	0	0	0
2	93	0	0	190,155	0	0	1
3	18	0	0	40,873	0	0	0
4	335	0	0	743,921	0	0	5
5	844	2	165	1,883,914	1.06	0.09	1
6	1,397	1	2	3,173,939	0.32	0.001	15
7	298	1	100	608,007	1.64	0.16	1
8	415	1	200	930,798	1.07	0.21	0
9	727	2	275	1,458,345	1.37	0.19	19
10	475	2	29	970,855	2.06	0.03	14
11	1,352	2	104	1,766,327	1.13	0.06	0
12	411	3	84	949,826	3.16	0.09	0
13	640	0	0	762,002	0	0	5
14	377	0	0	792,809	0	0	0

15	41	0	0	99,137	0	0	0
16	138	0	0	326,786	0	0	9
17	85	0	0	216,093	0	0	0
18	37	0	0	93,540	0	0	5
19	52	1	23	142,577	7.01	0.16	14
20	21	1	600	48,233	20.73	12.44	4
21	453	2	7	922,292	2.17	0.01	6
22	43	0	0	113,680	0	0	0
23	9	0	0	16,351	0	0	0
24	417	2	13	970,642	2.06	0.01	9
25	49	0	0	111,731	0	0	0
Subtotal	8,933	20	1,602	17,796,322			108
Average	357.3	.8	64.08	711,853	1.12	0.09	4.32
All Subcontractors							
1985							
Total	21,989	92	10,745	48,140,637			424
Average	239.0	1	116.8	523,268	1.91	0.22	4.61
1984							
Total	18,620	89	10,649	42,196,411			426
Average	N.A.	N.A.	N.A.	N.A.	2.11	0.25	

Source: Jidosha Friendship Association.
[a]Frequency rate = number of disabling injuries and illnesses per million hours worked.
[b]Severity rate = number of charges (in days) per million hours worked.

Table 4-4. Number of Injury and Illness Cases and Frequency Rate by Categories of Subcontractors and Year

Category	1982	1983	1984	1985
Machining				
Number of cases	71	49	37	32
Frequency rate	3.94	3.06	2.33	1.94
Stamping				
Number of cases	56	30	40	40
Frequency rate	4.51	2.46	3.16	2.88
Components				
Number of cases	28	25	12	20
Frequency rate	2.27	1.90	0.88	1.12

Source: Jidosha Friendship Association.

work injuries and illnesses per million hours in several firms to a high of 46.69 in one machining firm. As a group, firms making components had the lowest injury and illness frequency rate (1.12) in 1985, followed by machining firms (1.94) and then stamping firms (2.88). This same order occurred in the preceding year.

Severity rates ranged from a low of zero lost workdays per thousand labor hours in a number of firms to a high of 12.44 in one components firm. As a group, the components firms had the lowest injury and illness severity rates (.09) in 1985. They were trailed by machining subcontractors with .25 and stamping subcontractors with .36.

Compared with Jidosha, the subcontractors generally had inferior occupational safety and health practices and conditions. For example, during a brief tour of a subcontractor that makes seats for Jidosha vehicles, I observed a worker welding without either safety glasses or a shield. During a much longer period of observation at Jidosha, I never saw such an obvious breach of work-safety rules. The manager who accompanied me on the tour of the subcontractor did not even reprimand the worker. Further, there was confusion among managers about reporting requirements, particularly concerning the definition of a lost-work injury. This subcontractor was the largest of the ninety-two Jidosha subcontractors, and it had the lowest reported injury and illness rates. I had requested a tour of a smaller subcontractor, hoping to see firsthand some of the hazardous conditions other Americans who had visited such firms had described to me. The safety and health staff at Jidosha obviously did not want me to see such conditions and took me to this "showcase" subcontractor instead. The Jidosha safety and health manager who accompanied me on my visit was clearly embarrassed about the welding incident, however. This subcontractor had a reported frequency rate of .32 lost-work injuries and illnesses per million hours worked and .001 lost workdays per thousand labor hours in 1985, whereas Jidosha's rates were .47 and .03 on these respective dimensions for the same year. The subcontractor's rates must be viewed with suspicion because of the confusion regarding reporting and my observations of the work environment and the safety and health practices.

When it became clear that Jidosha safety and health managers would not agree to take me on a tour of a small subcontractor, one of my hosts, who was a semiretired senior executive with Jidosha, arranged a visit for me. Conditions at this subcontractor, a stamping firm, were much more hazardous than those at Jidosha or those at the other subcontractor I observed. Workers performing welding operations held in their hands small components that were being joined by machines that pressed and welded the components together. In most cases the workers activated the welder with a foot-control pedal or a one-hand control button. There were very few two-hand controls—a much safer system since one's hands are kept away from the pinch points. The reliance on workers to avoid accidents by being careful is characteristic of the Japanese approach to safety and health, but it was not working very well here. Managers conceded that accidents often occur because workers get their hands caught in the pinch points. The company's lost-work injury and illness frequency rate in 1985 was 11.63, more than four times the average for Jidosha stamping subcontractors and about twenty-five times the rate for Jidosha, and its severity rate was .21, compared to .36 for all stamping contractors and .03 for Jidosha.

Another obvious problem at this firm was that there was no system for ensuring that the air quality was adequate. There was a ventilation system that the managers said replaced the air in the plant with fresh air every thirty minutes, but they admitted that they did not know whether the air quality met recommended standards.

Both of the subcontractors I visited are first-tier subcontractors and are members of JFA. Conditions at lower-level subcontractors would undoubtedly be much worse. Furthermore, the differences in safety and health conditions in master employer firms and subcontractors are likely to be even more pronounced if the master employers are less vertically integrated than Jidosha. These companies are more inclined to contract out particularly hazardous work, such as stamping, welding, and repairing machinery. Jidosha contracts out some of this work, but it has considerable operations in each of these areas. The company even has a plant where it makes steel, one of the most hazardous stages of the automobile production process.

Cooperation and Conflict

The cooperative and conflictual aspects of labor-management-government relations at Jidosha are remarkably similar in virtually all important respects to other major Japanese automotive employers. Bilateral labor-management cooperation and loyalty are significant and highly beneficial. In particular, there is a great deal of cooperation in productivity- and quality-improvement activities. Production workers devote their complete attention and tremendous energy to their jobs. There is no horseplay; indeed, workers rarely smile or even talk to one another except about work-related matters. When the line stops for some reason, workers immediately begin cleaning around their work stations, performing maintenance on their tools and equipment, or engaging in other work-related activities. In contrast, workers in typical Big Three plants in the United States use down time for unscheduled breaks—to stand around joking, to smoke, to get a drink, or just to rest. Moreover, Jidosha workers work during the beginnings and ends of scheduled breaks and lunch periods.

Office workers at Jidosha demonstrate similar commitment. Approximately half stay on the job voluntarily after normal working hours. It is considered bad form and in a sense disloyal to leave work before one's immediate supervisor. Company officials insist that office workers have the *right* to overtime pay when they work late, but this right is rarely exercised. Nonetheless, the pace of work slackens noticeably after normal working hours, and office workers often remain at their desks simply to read newspapers. This behavior, which seems peculiar to Americans, reflects a variety of cultural characteristics related to both work and family expectations.

The company reflects its sense of responsibility and duty in its treatment of workers. Every effort is made to provide lifetime employment to permanent workers, even if this means "loaning" workers to Jidosha dealers, as was done in the aftermath of the energy crises in the 1970s, or moving around an underachieving worker until a job better suited for him or her can be found.

There has not been a strike at Jidosha since the mid–1970s, and there have been only four strikes in the union's forty-plus-year history (two of these strikes were for two days or less). As noted in chapter 2, one factor that enhances the harmony between labor and management is the mobility in the system, which offers union officials the potential to move into management. Although no union official would be so bold as to discuss such aspirations (especially with a foreign researcher), it is likely that officials with such goals work hard at minimizing conflicts and promoting cooperation. Moreover, former union officials currently in management contribute to this harmony. The union president spoke fondly of such former union members, whom he referred to as friends.

Several safety and health issues over which conflicts routinely arise in U.S.-owned plants are virtually conflict-free at Jidosha. For example, disagreements between line managers and safety and health staff over proper production procedures is not an issue here. Although line managers have far more status than equivalent-ranking safety and health staff, there are few disagreements because the Japanese typically have such respect for rules and regulations. If a procedure or facility is in violation, safety and health staffers can simply refer to government regulations or company policies to achieve compliance.

Another issue that is essentially conflict-free at Jidosha is the processing of workers' compensation claims. Conflicts arise in U.S. facilities (and indeed at Jidosha USA) because management suspects that a worker is faking or exaggerating an injury or illness to get a "paid vacation." The company may be concerned because it is self-insured for workers' compensation or because it fears its premiums will rise. Not one manager at Jidosha or at any of the other auto companies I visited in Japan ever complained about workers faking injuries or illnesses. Because of the social stigma associated with being suspected of such behavior, workers probably limit their workers' compensation claims more than any hard-nosed workers' compensation officer who screens claims of U.S. workers. Managers at Jidosha pointed out that workers' compensation was a nonissue for them since their injury and illness rates were so low that the company's premiums would not be affected by a few faked injuries or illnesses.

Although the Japanese safety and health system has a strongly cooperative orientation, conflicts among labor, management, and government are apparent on close examination. Labor and management have several points of disagreement. Naturally, these include wages, hours, and conditions of work that are negotiated in the collective bargaining system. At Jidosha, safety and health issues are not usually dealt with during normal labor-management negotiations. Typically, negotiations over these issues take place in safety and health committees, but these do not tend to be confrontational and considerable efforts are made to find compromises.

Safety and health managers at Jidosha contend that worker safety and health is a completely cooperative issue. Likewise, the union president initially contended that he did not even understand how labor-management conflict over safety and health could occur since both parties want improved safety and health conditions. These must be considered *tatemae* positions, however, since both the safety and health managers and the union president eventually conceded that there were conflicts, although they stressed that they were minor and temporary.

Company officials at Jidosha consistently display a high regard for government safety and health regulations and their enforcement. This is in marked contrast to views of OSHA held by both labor and management in the U.S. auto industry. The 1972 Industrial Safety and Health Act is pointed to as having brought about substantial improvements in safety and health in the Japanese work environment and for helping to lower the work injury and illness rate. Jidosha officials generally think the government legislation helped companies formalize their safety and health activities and improve their operation. Initially, Jidosha safety and health managers also expressed favorable opinions about the administrators of the state-plan OSHA with whom they had been in contact regarding the start-up of Jidosha USA. It became clear, however, that these relations were not nearly as cordial as they were with the Labor Standards Bureau, and they deteriorated noticeably after production began.

The Japanese government relies on master employers to help improve the work environment in subcontractor firms, indicating further cooperation between government and industry. Informal contacts between government and Jidosha also allow questions to be asked and suggestions to be made without concern about the repercussions. Government safety and health inspectors with the Labor Standards Bureau on occasion ask the advice of Jidosha safety and health experts on the handling of safety and health problems at Jidosha's subcontractors. Given the nature of regulatory relations and master employer–subcontractor relations in Japan, however, Jidosha officials are expected to do more than passively provide advice; they are expected to solve the problems.

Several upper-level Jidosha safety and health managers have also been appointed members of special Labor Standards Bureau committees, such as the Stamping Accident Prevention Committee on which the general supervisor of the Safety and Work Environment Division serves. Members of this committee serve as ad hoc inspectors for the Labor Standards Bureau by providing advice to smaller firms in the region, especially subcontractors of the committee members' firms. Such members are said to be "on loan" to this committee for three or four days each year.

One anecdote regarding safety and health regulatory relations is noteworthy. An official of the Labor Standards Bureau was a guest at a Jidosha safety awards ceremony when he noticed a safety hazard in the plant. Unofficially, he informed the company safety and health staff about it, and they had the problem corrected. There were no repercussions, and the matter ended there.

Jidosha was inspected in 1982 and again in 1985. Company officials thought the inspections were beneficial but admitted to having some anxiety in anticipation of them. (As is standard procedure in Japan, company officials were notified about the inspections.) Such anxiety is to be expected. Even if no serious violations were cited, a poor showing would involve considerable loss of face for the safety and health staff and the responsible line managers.

In the 1985 inspection twelve Labor Standards Bureau inspectors found a total of nineteen hazardous conditions or organizational problems. These included inadequate guards on or around some machinery (three cases), insufficient ventilation or dust collection or excessive noise in certain work areas (seven cases), inadequate or missing signs around some hazardous machines (two cases), not holding certain safety and health meetings at the prescribed frequency interval (one case), workers not wearing the proper safety and health equipment or following proper procedures (three cases), and inadquate clarity of responsibilities and authority of safety and health staff vis-à-vis line managers (two cases). The company was also asked to improve its system of inspecting machinery for hazards. No fines were imposed, the company agreed to correct all the problems, and it promptly did so. Jidosha officials thought the safety and health conditions in the plant improved as a result of the inspection.

Reflective of the cordial—some would say excessively cordial or cooptive—regulatory relations in Japan, it is common practice for company officials to take Labor Standards Bureau inspectors out to lunch or dinner and drinks after an inspection (Kamata 1986). Jidosha officials, no doubt, followed this custom.

When questioned about the relation of safety and health to productivity and profitability, the initial and persistent response from management, government, and union officials was that enhanced safety and health measures *always* promote greater productivity and profit. Unwilling to accept this assertion at face value, I posed several counterexamples. Finally, the safety and health managers conceded that there might be exceptions to the rule. They acknowledged that in some cases a large investment to make a modest improvement might not be cost effective and would therefore not be undertaken. The specific example cited was air-conditioning the plant to alleviate the oppressively hot and humid conditions during the summer. One worker complained that the temperature sometimes reaches 40°C (104°F).[14] The foundry shop, where steel is heated up to 350°C (662°F) is hotter still. The cost of air-conditioning the whole plant, which was old and would have required retrofitting, was considered prohibitive. As an alternative, spot coolers are used. Workers still complain, arguing that the spot coolers blow air into a limited area and their jobs require them to move around to areas that are not cooled.

Local building requirements contribute to this problem. Because one side of the plant faces a neighborhood whose residents would be distracted by noise and light from the plant, it has no windows on that side. This cuts down on the air flow in the summer, exacerbating the level of discomfort.

In general, it seems that the high costs of constructing new plants, stemming from the extraordinarily high costs of land, result in compromises in safety and health. Workers and managers need to learn to adapt to older, and in some instances less safe, factories that might have been replaced if the cost were not so high. Although most Jidosha production facilities I visited were in good condition, some had safety or health problems because of their age. The number of Labor Standards Bureau recommendations to improve ventilation or dust collection at Jidosha attests to this. Further, a senior executive working in Japan with Jidosha's Big Three American partner spoke disparagingly of an old Jidosha engine plant that uses an oil bath to cool the engines. He said the plant has a permanent blue haze from the atomized oil particles in the air. I was probably steered away from this plant on purpose.

Jidosha officials concede that the cost of complying with government regulations is a problem, especially for smaller companies. As one official put it, "I am afraid that generally small companies may suffer from Japanese legislation because it is very difficult for them to meet the legislation that requires/costs a lot of investment, so Japanese small enterprises are suffering. We don't have any conflict with the Japanese government."

Production managers admitted that some government safety and health requirements do compromise quality. A manager at the Jidosha subcontractor that manufactures seats said that government requirements that fumes be ventilated rapidly in welding operations lowered the quality of the welds. As a result, the subcontractor is investing in robot welders. As another example, a company official said that noise buffering on stamping presses can reduce the quality of the components that are pressed.

Temporary workers are a source of labor-management conflict. The union president noted that the union opposes the hiring of temporary workers for safety reasons and feels that those who are hired should be better trained. His preference is to increase the permanent work force.

Another issue that demonstrates both the cooperative and conflictual aspects of safety and health is the practice among some workers of working during part of their breaks or lunchtimes. One reason for doing this, of course, is to keep pace with production. No manager or worker complained about this practice. Rather, the concern was that it occasionally leads to accidents, in the union president's words, "resulting from good will."

Managers conceded that occasionally workers do not follow proper safety and health procedures, such as wearing safety hats, safety glasses, and respirators, and need to be disciplined. As noted earlier, when compared with U.S. firms, such occurrences are relatively rare in Jidosha and in Japanese auto companies in general. Further, in contrast to the United States, there does not seem to be much overt controversy concerning the discipline when violations do occur. In Japan, the union, co-workers, and everyone else—except perhaps the worker being disciplined—seem to look on discipline in such cases as appropriate. On close investigation, however, the form of discipline used is a sensitive issue.

The methods of discipline used are closely related to the Japanese sense of face or honor. The most common practice is for the supervising manager or foreman to verbally reprimand the disobedient worker. In evaluating the effectiveness of this practice, a line manager stated:

> My action is very effective to the person who violates rules and other operators who are working with him who violate the rules. In addition, that depends on how I should scold him and also I suppose the final goal is that each operator has to be conscious about each other. Sometimes it may be difficult because of [the desire for] good human relationships to be kept in workshops. So we encourage or we try to scold him who violates the rule, and that is very effective because of the sense of shame about violating rules and also because workers are very committed to the goals of the organization such as the zero-accident goal.

Should the manager wish, he can put the reprimand in writing. Workers who receive such reprimands usually receive a lower performance rating at their evaluation. Although there is no specific performance category for safety and health, there is a category that includes following work rules. In most cases, receiving a low rating would be regarded as a considerable disgrace.

Some managers have gone much further than this. One manager made a worker come to his office for a scolding and to write an apology and a promise that he would not violate work rules in the future. Another worker in a different department was required to apologize to his fellow work-group members for not wearing the proper safety and health equipment. Yet another was dismissed from his normal production activities for two hours and required to stand beside the department manager. The following day he was permitted to go back to his job but was required to wear a yellow shirt designating him as a rule breaker. The department manager felt that this discipline was very effective in discouraging a recurrence of the violation and in dissuading other workers from such behavior.

Safety and health officials were demonstrably upset that I learned of the latter method of discipline, and they requested that reference to it be deleted from a report on my research that was being given to top management. The officials felt that it was an extreme form of punishment and contended that it had been used only in very special cases in the past and was used even less frequently now.[15] One safety and health official commented:

> I think that this work practice, even if it was effective, this kind is a little excessive. Of course that really depends on the degree of violation. If a manager gave [a] caution to the rule breaker one hundred times, if he still breaks the rules this may work very well; but without giving any caution and that manager did this way, that manager was too excessive.

In explaining the severity of this punishment and the sensitivity about it, a safety and health official referred to a Samurai tradition: "In ancient times, Samurai killed himself if he failed in his job. They were allowed to kill themselves. That is a type of honor rather than continuing to live with shame."

In response to further questioning, the officials affirmed the implication of this statement—that the worker disciplined in this way probably would have preferred suicide to the discipline. Company officials made a point of noting that no company employee disciplined in this way had ever actually committed suicide.

After reading about the incident in my report to management, the union president commented that he was not aware of it but that he would investigate the matter. He felt the discipline was extreme.

Occasionally, the union is drawn into safety and health conflicts. It attempts to prioritize workers' requests for improvements in the work environment, for example, and inevitably upsets workers whose requests receive low priority. According to the union president,

> We may have to give priority to some of those so we may not be able to present everything that is put forth to us by the union members, but we have to view it from most important on down. So those union members who have been put on the lower priority list may continue to have complaints and may be less sensitive to the difficulty.

In recent years, the union has also reprimanded workers who attempted to conceal injuries or illnesses or were engaging in other safety and health violations. The union president complained that union members who work when they are sick or injured are sometimes a "headache to the union." He said that the pursuit of the zero-accident goal could exacerbate this problem, thus "defeating the very objective of this zero-accident promotion." He knew of two specific cases in which workers concealed injuries and, in his words, "In both cases we took very strong measures to give proper guidance."

Workers were much quicker than managers or union officials to point out conflicts, although none was viewed as extraordinarily contentious. Several workers mentioned that they had complained to management about the heat in the summer and the cold in the winter. They conceded that spot coolers had helped during the summer but they did not consider them an adequate solution to the problem.

A related concern was the level of dust and fumes in the workplace. The workers favored greater use of "dust collectors" or improved ventilation systems so that they did not need to wear masks. One team leader who complained about the level of dust generated by a sanding machine in his work area told of his unsuccessful efforts to get a sophisticated ventilation system used elsewhere in the plant placed in his work area: "I am making a proposal to use this type of machine . . . but management doesn't accept my proposal because of limited budget." One worker noted that masks had come into widespread use only within the last five or six years. (Management contended, however, that workers were first required to wear them about ten years before that.)

Another worker mentioned that some managers did not obey the traffic rules on the company's premises. He said they had "high pride" and that the behavior sometimes angered him. Obviously, such managers had not been impressed by the "Patiently, generously and tolerantly wait" traffic slogan.

Another issue that workers did not mention initially but conceded was a problem when asked about it was the pressure not to report injuries or illnesses. There was a consensus among several workers that workers would report major injuries or that co-workers or the foreman would notice the accident as soon as it happened. Minor injuries or illnesses were a different matter. Workers indicated that when an accident was their own fault and no one else noticed an injury, they would be inclined not to report it for fear of looking stupid and of letting down their colleagues. One worker expressed concern that others would laugh at and ridicule him if he reported a minor injury that was the result of his own negligence. They noted, however, that in the past some workers had even attempted to hide major injuries. By placing such emphasis on predicting and preventing hazards, Jidosha may actually make the problem of underreporting worse. If workers are convinced that the work environment is safe and healthful and that accidents are their own fault, they may be less inclined to report injuries and illnesses.

One other area of conflict concerns the filing of grievances. Grievances are a sensitive issue with management, and Jidosha officials are unwilling to provide data on the number or types of safety and health grievances that are filed. No doubt management fears that the grievances would contradict their *tatemae* descriptions of labor-management relations. I was told the grievances were related to workers' requests for improvements in safety and health conditions beyond what is legally required, which makes sense. If the grievance pertained to violations of safety and health regulations, the dissatisfied worker could request a government inspection, with reasonable prospects of maintaining his or her anonymity. The worker filing a grievance with the company is immediately identified and considered a troublemaker.

Conflicts also arise over personal vacation days, the amount of overtime scheduled, and the pace of work. Although safety and health staff and the union have been trying to encourage workers to use more of their personal vacation days, some workers, especially those in higher-skilled jobs, said that they felt pressure from their foremen as well as personal (internal) conflict because of their work-group loyalty that inhibited them from using these days. They felt they were needed too much to take a day off. In one case, a quality-control inspector indicated that this pressure led to physical and nervous exhaustion. He felt a tremendous sense of responsibility to ensure that there were no defects in the cars he approved. Moreover, he felt he needed to be at work every single day because there was no one to replace him. He indicated that the situation looked more promising because Jidosha had begun its multiskilled worker program. When fully implemented, it would be easier to replace him when he took vacation days or was sick or injured.

The conflicts over the length of the workweek and the amount of scheduled overtime have important safety and health dimensions, similar to those regarding the use of vacation days. That these issues have been heatedly debated in national and even international forums is evidence of their significance.

The pace of work at Jidosha and at other Japanese auto plants is considerably faster than at U.S. plants, as reflected in the higher productivity of Japanese firms; however, no one complained about the pace. To do so would probably be viewed as an admission of weakness in front of co-workers and therefore a loss of face. Workers noted, however, that approximately 10 percent of freshmen (new hires) have difficulty keeping up. The Japanese investigative reporter Satoshi Kamata (1982, 1986) described complaints from even seasoned workers. Kamata worked under a pseudonym as a seasonal worker in a Toyota assembly plant to investigate working conditions. Moreover, the pace of work and the amount of compulsory overtime and holiday work were targets of complaints by a large majority of Japanese autoworkers surveyed by the Federation of Japanese Automobile Workers' Unions in 1990. These practices were blamed on the very short cycle for introducing new cars in Japan (four years compared with six to ten in the United States). The federation requested that the manufacturers lengthen the cycle to reduce work pressures (Solo 1990; *Wall Street Journal* 1990).

It is not unusual to see workers literally running to keep pace with production. Thus, starting work early or working partway through a break may be more than a matter of loyalty; it may be necessary to keep up. A delegation of UAW officials visiting another Japanese automaker remarked that the workers looked "like breakdancers, only faster" (Kamata 1986). Moreover, *kaizen* activities seek to use 100 percent of the worker's time so that no production time is lost. The housekeeping and maintenance activities required during down time are further examples of ways Jidosha tries to ensure that workers waste no time. Some labor-oriented critics of Japanese management practices complain that before-work and after-work meetings and evening socializing in which the only subject of conversation is company business are attempts to get more than 100 percent of the workers' time.

Production workers often speak of the rhythm or harmony of the work process, but without a doubt it is physically and mentally exhausting. Moreover, stress-related injuries and illnesses are likely to result, as well as injuries from rushing. There is also the temptation to shortcut safety and health measures and procedures thought to inhibit speed. Some Japanese workers give in to this temptation, although, because they have greater respect for rules and regulations, they are less likely to than American workers.

Certain jobs also create safety and health problems because of the exertion required. Again, no worker complained about this, but an assembly plant manager conceded it was a problem when he responded to a question regarding what he would do with respect to safety and health if he had a 50 percent increase in his budget: "We'd like to use the additional budget for the operations which require heavy loads such as loading batteries or loading seats. The operator requires heavy loads, so these workshops would like to take measures so that they can ease work so that they can be released from such loads." He also said he would address the problem of heat.

There was little evidence at Jidosha that labor is dissatisfied with government safety and health regulations. On the whole, union officials and workers were satisfied. Local Labor Standards Bureau officials noted, however, that occasionally, just before elections, the bureau has received complaints about the safety and health conditions in local companies. Given Jidosha's predominance in the local economy, Jidosha or one of its subcontractors has probably been the subject of some of these grievances. The feeling is that the workers making the complaints are attempting to discredit their employers and indirectly discredit the ruling Liberal Democratic party, which most large corporations support. The complaints are investigated, although officials are sometimes skeptical about their legitimacy. Compared with the United States, such incidents are rare. In small businesses especially, workers are apparently fearful that they could easily be identified and, although it is prohibited by the Industrial Safety and Health Law, that they could be subject to retribution.

The president of the Jidosha Workers' Union noted that in the past when an accident occurred the union demanded that the machine involved be shut down. Management, in response, would argue that this would "bankrupt the company." More recently, the union and management have become more cooperative in seeking to eliminate hazards in ways that do not compromise production or involve excessive costs. State-plan OSHA officials who visited Jidosha facilities in Japan were surprised by the extent to which the union was concerned about production. On several occasions they suggested modifying machines or production processes to protect workers further but were rebuffed by union officials who said the changes would hurt production.

Many of the differences in labor-management-governement relations between the United States and Japan are summarized in two popular sayings. In the United States, one is told that, "the squeaky wheel gets the grease," that one must be assertive and, in some cases, confrontational to get what one wants. There is no such saying in Japanese. The Japanese are told that "the nail which sticks out gets hammered in," meaning that conformity and harmony are desired attributes and that actions that disrupt the harmony will be punished.

In the next chapter I look at the way in which Jidosha's safety and health management system has been adapted at Jidosha USA. Given the enormous differences between Japanese and American cultures, work practices, and styles of worker protection, this has been a great challenge for both the Japanese and Americans at Jidosha.

5

Occupational Safety and Health at Jidosha USA

Jidosha began production operations in the United States in the latter half of the 1980s with the establishment of an assembly and manufacturing facility. The decision to start U.S. operations was a major gamble forced on the company. Under the import agreement between the United States and Japan, the Japanese government had allocated shares of Japan's auto exports to the United States based on export figures for a period when Jidosha's performance was weak. Moreover, U.S. output was not subject to the import agreement, and by the time Jidosha decided to construct its facility, several other Japanese automakers were already producing vehicles in the United States. The slide of the dollar relative to the yen made U.S. production even more attractive. It is not clear whether managers at Jidosha anticipated this decline or whether they were just lucky; in either case, the exchange rates at the end of the 1980s made establishing U.S. operations a good decision.

The Jidosha USA plant employs a total of about three thousand managers, production workers, and support personnel and has a projected capacity of approximately 240,000 automobiles per year. Productivity at full-strength employment and full capacity is about 69 vehicles per person. This might appear high relative to levels in Japan (approximately 49 vehicles per worker), but it can be explained by the differences in vertical integration. Jidosha USA is primarily an assembly plant, whereas Jidosha operations in Japan comprise virtually all stages of the production process. Large numbers of corporate personnel not directly involved with production are included in the calculations for Jidosha in Japan. Nevertheless, productivity at Jidosha USA is high compared with U.S.-owned assembly plants. One Safety and Health Department staff member observed that the rate was more than twice that of the Big Three plant where he had worked.

As of this writing, two models, virtually clones of each other, are being produced at the U.S. plant, and production is about to begin on a third. Both current models are midsized sports sedans that have proven to be favorites with consumers. One is marketed by the American partner of Jidosha, which cooperated on its design and supplies various components. Some of the models produced by the American partner are being exported to Japan.

The plant uses state-of-the-art technology and employs more than 350 robots. Many sport such nicknames as Robert DeNiro, Joanne Woodward, Steve McQueen, Jim Brown, and Daffey Duck.

The facility is modeled after the relatively new and modern Japanese Jidosha facility, located away from the main plant. The U.S. plant is extremely clean, well lighted in most work areas, and quiet relative to American-owned plants and the main Jidosha facility. There is considerable green space in the area surrounding the plant and even a picturesque Japanese garden in a courtyard.

On one side of the garden, in the center of the plant, are the administrative offices. They are within easy access of all parts of the plant; office workers and production workers can get to them without wasting time or effort. The cafeteria is across the courtyard from the administrative offices and is also accessible for both production and office workers. In addition to being a place to eat, the cafeteria is used for workgroup meetings and as a place where workers can read or study.

Economic conditions have been depressed in the state and region where the plant is located, so that the local community, for the most part, was delighted that Jidosha chose to locate there. Although a holiday from local taxes used to attract Jidosha has proved to be controversial, residents of the community note that Jidosha has been a boon to employment. Supplier firms have also sprung up, including several Japanese subcontractors that followed Jidosha to the United States. They employ even more people and inject more income into the local economy. Moreover, Jidosha is considering investing in an engine plant near its facility. Residents of the community are quick to say that Jidosha is a good neighbor, citing the cleanliness of the operations and the plant's aesthetically pleasing design and grounds.

Production at Jidosha USA is similar to production at Jidosha plants in Japan, especially the new facility on which the U.S. plant is patterned. The team concept of production is used, as well as a just-in-time inventory system. Parts and raw materials are bought from both Japanese and American suppliers. The exact combination is a sensitive point, however, given that Japanese automotive firms have been criticized for having insufficient domestic parts in the vehicles they produce in the

United States. (As a further example of this sensitivity, Jidosha officials insist that the facility not be called an assembly plant. They prefer to refer to it as a "manufacturing facility," which suggests greater value added in the operations.) Only one vendor brings components onto the factory floor. The others deposit their wares in the Materials Handling Department. This contrasts with Japanese plants, where vendors routinely enter work areas and control of their activities constitutes a safety problem.

Because Jidosha USA is a separate corporation consisting of the one plant, both the plant and the corporation have parallel structures. In most cases the same individual fills both the corporate role and the plant role. The president of the company as well as his two executive vice-presidents (for administration and for operations) are Japanese. Under these managers are the vice-president of personnel, who reports to the executive vice-president of administration, and the vice-president of manufacturing, who reports to the executive vice-president of operations. Reporting to the vice-president of manufacturing are the managers of the various operational departments: Paint, Body Assembly, Stamping, Trim and Final Assembly, Plastic, Materials Handling, Quality Control, and Maintenance. Assisting (some would say overseeing) the managers in each of these departments are Japanese advisers. They report to a senior Japanese adviser who, in turn, reports to the executive vice-president of manufacturing. Reporting to the vice-president of personnel are several "directors" (equivalent in rank to managers) of functional departments, including Safety and Health.

Reflecting the belief held (or at least professed) at the start of operations that American managers know better than Japanese how to deal with American workers, most personnel directors do not have Japanese advisers. The one exception is the safety and health director. The appointment of a Japanese adviser reflects the importance attached to this function, the confidence Jidosha management has in its safety and health program, and the fact that the safety and health function is viewed as having both operational and personnel dimensions. Because operations at Jidosha USA were started only recently, the company's structure, staffing, and personnel policies are still in flux. This chapter concentrates on conditions as they were one to two years into production.

Labor Practices

The work force is composed of approximately 2,550 direct workers (i.e., production workers) and 450 indirect workers, including managers.

These figures do not include temporary Japanese advisers because the number fluctuates widely. Total Japanese staff at the plant was as high as 500 at the start of operations but is now much lower.

Subsequent to the time period on which I am focusing, Jidosha hired several hundred more American workers, many on a part-time basis, to reach full-strength employment of about 3,500. Part-time workers are used in combination with a practice of contracting out some activities as a way to protect the jobs of permanent employees. During a downturn, the part-timers can be laid off and the subcontracted work taken over by regular employees.

The status of the part-timers proved to be a major source of contention. Both the workers and the local union, of which they were members, protested that they were being treated as second-class citizens, in that they received lower wages and few benefits, even though they sometimes worked forty hours a week. As a result, the company agreed to hire many of them on a full-time basis. Others were laid off when demand for Jidosha vehicles fell off in the third year of operations. After the layoffs, part-time employees totaled less than one hundred, but their status has continued to be a source of friction both within the union and between the company and the union.

The stated intention of Japanese Jidosha management was to develop a "third style" of management at Jidosha USA incorporating the best of Japanese and American techniques. One of the coordinators of the training program in Japan for Jidosha USA personnel stated:

> We at Jidosha do not request, strongly request, them to follow Jidosha's corporate culture. But we are aiming for, our goal is that, Jidosha USA people can find a third way which is different from American-style, which is different from Japanese-style management—the third way. The third way can be realized by the good combination between the American-style management and the Japanese-style management. . . . If we try to apply Japanese-style management to the American style, we would have very much difficulty. So what we have now considered is that we are going to adopt the new teamwork that is different from the Japanese style because this new teamwork concept is based on the individual. Individuality is incorporated into this teamwork policy. Generally in Japanese style, teamwork is based on groupism. . . . Individuals are likely to be sacrificed by group work. That is Japanese style. So what we are considering is the new teamwork concept that is based on the individuals.

As we shall see, Jidosha has had some trouble developing this style. The individuality of American workers has proven to be more incompatible with the Japanese management style than Jidosha managers had anticipated, and Japanese managers have had difficulty letting go of their group orientation.

Jidosha USA follows many of the now-familiar personnel policies of Japanese employers that seek to break down barriers between management and hourly workers. For example, there are no reserved parking spaces for senior executives; everyone eats in the company cafeteria; and everyone, including the president, wears the required Jidosha USA uniform.

Media reports indicated that management also agreed to take pay cuts before laying off workers, but lower-level U.S. managers I spoke to did not know of the existence of this provision in the labor-management contract. The safety and health director was unaware of it and initially claimed that no such agreement existed. When he was shown a magazine article that referred to it, he was surprised and remarked that he and other U.S. managers would be very upset if these provisions were ever enacted.

Typical of Japanese production processes, workers at Jidosha USA have considerable responsibility. They are expected to inspect the work done by those who preceded them on the production line, and, as in Japan, they are able to signal supervisors or stop the line if they discover a problem. This system is in keeping, of course, with the Japanese manufacturing philosophy of *jidōka*, bringing a problem to everyone's attention so that it can be resolved immediately. There have been reports, however, of both subtle and overt pressures on workers to avoid disrupting production unless a problem is very serious. The pressure can be particularly severe for probationary employees, who have difficulty keeping pace with production.

Like Jidosha in Japan but unlike other U.S. auto firms, there are very few worker classifications here. There is only one category of production worker and one category of skilled craft worker. This classification scheme promotes flexibility and efficiency but is typically resisted by workers, who fear a loss of job status, seniority rights, and other benefits of the traditional system (Parker and Slaughter 1988). In this case, the union felt it needed to concede on this issue to attract Jidosha to the area.

The organization of work groups is similar to that in Japan. Workers are organized in teams of approximately four to ten workers each. Team leaders, who are production workers and members of the union, fulfill a number of recordkeeping and planning functions normally filled by foremen in traditional U.S. plants. First-line supervisors, comparable to foremen in U.S. plants, serve as unit leaders and are responsible for two to four teams. As in Jidosha in Japan, workers are expected to learn the jobs of all other members of the work team.

Jobs at this plant, as at other Japanese transplants, are considered highly desirable, and management has been able to be very selective in its hiring decisions. More than 100,000 applications were received for 3,500 positions, representing approximately a thirty-to-one applicant-to-job ratio. The application process consists of five steps, each requiring a trip to the test site. Applicants undergo interviews; paper and pencil tests to measure basic intelligence, problem-solving skills, listening skills, and mechanical skills; physical examinations that test for basic health as well as alcohol and drug use; and two separate examinations at the assessment center, one that evaluates attitudes conducive to cooperation and teamwork and another that measures mechanical ability. Such examinations are extremely rare in the United States for selecting production workers. They are most commonly used in filling managerial positions.

Unlike most other Japanese transplants, which fought unionization efforts, Jidosha signed a letter of intent early on, agreeing to allow the UAW to organize the plant. Unconfirmed media reports indicate that Jidosha's management had reservations about the plant's being unionized but that the company's Big Three partner prevailed upon it to agree to unionization in the hope that its own laid-off union workers would be hired there. Other Japanese transplants have been accused by the UAW of screening out applicants who have a history of being in a union or who are judged to have union sympathies. It is not clear whether unemployed UAW members were actually given preference in hiring decisions. On the one hand, their work experience might have been considered in their favor. On the other hand, they may have been assumed to have the poor work habits and confrontational approach to labor-management relations perceived to be common among U.S. auto workers. In fact, a considerable number of the workers Jidosha hired had never worked in the auto industry or in any manufacturing job. The chairman of the union at the plant remarked that many workers were "flipping burgers at Burger King" before starting at Jidosha.

Some would argue that the careful screening process backfired to some extent. Several disgruntled employees have been very articulate and savvy in making their complaints public and in pursuing them through grievance procedures and legal forums. Their problems have proven to be a major headache for Jidosha USA management and at least *seem* to be more serious than at other Japanese transplants. Worker discontent may actually be no more serious. Rather, the union may provide protection to workers who want to go public with their complaints or who are willing to talk about them to researchers. (In contrast, one needs to develop a very high degree of trust before personnel at other Japanese transplants will feel free to criticize their employer.) Given the Japanese sense of "face" and their abhorrence of bad publicity, it is likely that many in Jidosha's management regret the decision to accept unionization.

The compensation system for hourly workers is fairly typical of the one used at U.S. auto plants and therefore is very different from the one used at Jidosha in Japan. Wages for hourly workers are somewhat below those negotiated by UAW workers with the Big Three automakers, but are scheduled to match theirs by the end of the current contract, which expires in the early 1990s. Team leaders are paid a modest hourly premium. There is also a bonus system for hourly workers who have perfect attendance.

There are two bonus plans for management: SOAR and TOPS. SOAR (Successful Objectives Achieve Rewards) provides bonuses of up to 22½ percent of base salary to managers who successfully achieve specified goals in the year. TOPS (Team-Oriented Production System) provides bonuses of up to 7½ percent per year based on "soft" or interpersonal skills. Components include performance of *kaizen* activities, team orientation, flexibility, treatment of other people, and community involvement. Efforts are under way to devise a bonus plan similar to TOPS for hourly workers. Such a system would represent a significant departure from the compensation system used in the rest of the U.S. auto industry.

As is typical of U.S. plants, Jidosha USA has a progressive discipline system for violations of the company "code of conduct." It is distinctive, however, in that in this five-step system the step before dismissal is one day off *with pay*, designed to give workers time to think about their offensive behavior and develop a plan of corrective action. This non-punitive approach is more in keeping with Japanese traditions of discipline than American traditions. The overt use of shame was not nearly as pronounced as in Japan, but there was plenty of evidence of peer pressure, particularly if someone did not pull his or her full share of the load and other members of the work team needed to pick up the slack.

Several other aspects of the code of conduct are more consistent with work practices in plants in Japan than in the United States. These include prohibitions on eating, drinking, smoking, or having reading materials except in designated areas. These prohibitions have implications for productivity and quality as well as safety and health.

A grievance procedure is also in place. In keeping with the attempt to establish a positive orientation to labor-management relations, it is referred to with the upbeat title of the "problem resolution procedure."

Not surprisingly, workers at the U.S. plant are much more willing to verbalize their complaints with working conditions and to use grievance procedures than are their counterparts in Japan.

The presence of the UAW is significant in labor relations at this plant. In many ways, it appears that Jidosha USA is attempting to involve the local union as a partner in management, similar to the way the Jidosha union operates in Japan. Union officials hold full-time jobs with the company in the Department of Personnel and the Department of Labor Relations, where their desks are located. They are paid by the company, but they may work on union activities at their own discretion. It is highly unlikely, however, that anything approaching the highly cooperative union-management relations prevalent in Japan can be achieved here. The election of a slate of candidates affiliated with the New Directions movement in a plantwide election held a year and a half after the facility opened supports this view. New Directions has opposed many aspects of labor-management cooperation and has accused the UAW of "selling out" to management. The success of its slate was attributed to worker discontent and disillusionment.

Total absenteeism at the plant runs about 4 percent a year, compared with 3.3 percent at Jidosha in Japan. The figure for most U.S. auto plants tends to be higher.[1] Because of the busy schedule, workers at the American plant, like those at Jidosha in Japan, do not use all of their vacation days. One significant difference, however, is that whereas workers in Japan have no financial incentive to do this, production workers in the United States receive an extra day's compensation for each vacation day they work.

With the attendance bonus system, hourly workers with perfect attendance can earn as much $500 extra per year. Such a system would be considered odd in Japan. The Japanese would wonder why one should pay a worker extra to fulfill his or her duty; he or she is being paid already and has a duty to come to work. Increasingly, American workers have been using more of their vacation days, leading some Japanese to question their loyalty to the firm.

Two years into operation, the turnover rate at Jidosha USA was about 3 percent per year. This was higher than the rate for Jidosha in Japan (2.4 percent) but lower than the estimated rate for the rest of the U.S. auto industry (5 percent).[2] It is not surprising that the turnover in the United States exceeded that in Japan, especially since the U.S. operations were still in the start-up phase. Many of the personnel practices, though commonplace for Japanese transplants, were new to these workers, and some people could not adapt to them.

There is concern at Jidosha USA about the number of top-level American personnel who have resigned, including three labor relations directors in the first three years of operation. A common complaint among these managers is that they had little autonomy and most decisions were made in Japan or by Japanese managers at Jidosha USA. American managers feel that there is a limit to how far they can advance in the company given the apparent desire to keep control in Japanese hands. Managers have difficulty adapting to the consensus style of decision making in those areas in which they do have influence, further eroding their autonomy. In certain respects, American managers have had more difficulty adapting to the Japanese management system than have American production workers. Many production workers have found Japanese managers to be more cooperative than their American counterparts.

Approximately 29 percent of the workers at this facility are women, many of whom work in production jobs. This figure is high for a U.S. auto plant but even more impressive when compared with the figures for other Jidosha plants or for other Japanese automakers (in Japan), where there are no women in production jobs. Furthermore, women have held director-level positions in the Personnel Department. Undoubtedly, the Japanese managers of Jidosha were sensitive to international criticism and legal problems in the United States regarding the treatment of women by Japanese companies. For this reason, they seem to have made an effort to be fair in their hiring and treatment of women. Clearly, though, long-held, deeply ingrained attitudes toward women cannot be changed overnight, and some women at the plant have complained about their treatment by the Japanese. These problems have not deteriorated to the litigation stage, however, as they have at other Japanese employers in the United States. Responsibility for the hiring of women and for other personnel issues has been handled by American Personnel Department staff, who were among the first Americans hired by Jidosha USA.

Minority employment is also a sensitive issue. Approximately 17 percent of the work force are minority group members, of which the overwhelming majority are black. The percentage of blacks in the surrounding community is somewhat higher (about 29 percent), but the percentage at Jidosha USA is probably higher than at any other Japanese automotive transplant.

Sensitivity to minority group issues was heightened by the negative publicity Japanese companies in the United States received following former prime minister Nakasone's now-infamous comments about the intelligence level of Americans. (He said that one of the reasons for the relative success of the Japanese was that they were smarter than Americans, in part because of the racial and cultural homogeneity of Japan compared with the heterogeneity of the United States.) Shortly after this statement was reported, presidential candidate Jesse Jackson went to Japan to meet individually with the presidents of leading Japanese firms doing business in the United States, including Jidosha. Jidosha's top management was quite concerned about this meeting, and its international director of public relations solicited my opinions regarding Jackson's likely demands and reactions to Jidosha's possible responses. While in Japan, Jackson demanded greater reciprocity in economic relations between Japanese companies and the American black community and threatened to boycott Japanese products if improvements were not forthcoming. Sensitivity to racial issues was further heightened by media reports of discrimination and even violence in the United States directed toward Asians and toward Japanese in particular.

Jidosha USA seems to have made serious efforts to head off problems in race relations. There were no obvious Japanese-black racial problems when I visited, although Japanese personnel may well have been repressing (or at least not expressing) their feelings of racial superiority. It is also likely that Japanese-black relations have been facilitated by the screening process, which emphasizes selecting people who could get along with others in groups. Moreover, it is reasonable to assume (or at least to hope) that having an opportunity to work and interact with each other has had a liberalizing effect on the racial attitudes held by both the Japanese and blacks.

There was some evidence of black-white tension. Bathroom graffiti, such as one comment accusing blacks of being lazy and not pulling their weight in work groups, was indicative of such tension. There was also some indication of minor tensions between the Japanese and whites. Overall, however, racial problems were probably no more severe than one would find in any manufacturing facility in a similar location. Jidosha is one of very few Japanese automotive plants to locate in a community with a high concentration of minority group members. Japanese automotive companies have come under strong criticism for demonstrating a pattern of avoiding such areas (Cole and Deskins 1988).

There are two shifts at the Jidosha USA plant, the first from 7:00 A.M. to 3:30 P.M. and the second from 7:00 P.M. to 3:30 A.M. Managerial personnel sometimes work split shifts to provide coordination. The stated reason for this somewhat unusual schedule is to enable workers to spend time with their families even if other family members have to go to work or school. Workers are assigned to one shift (some degree of choice is available based on seniority), rather than alternating shifts weekly as is done in Japan. Neither shift is viewed as superior, and people from both shifts have bumped their way onto the other. Workers on the night shift receive a premium, however.

The well-known Japanese practice of having morning exercises was initially well accepted at the U.S. plant. During the start-up phase of operations, approximately 90 percent of the workers participated. These loosening up exercises, which last only about ten minutes, have safety and health benefits since they reduce the risk of muscle pulls and strains just as stretching exercises before competition or practice reduce an athlete's risk of injury. Two years into operation, however, very few workers still do the exercises, and safety and health officials think this may be contributing to injuries and CTDs at the plant. There has also been a decline in workers' enthusiasm for the Japanese style of management that extends far beyond this activity.

Labor-management relations at Jidosha USA are entering a crucial phase. In effect, the honeymoon is over. Many workers feel they were misled into believing that the plant would be a workers' paradise where the interests of workers would be paramount. Many feel they were misled when they relied on management assurances that Jidosha USA would combine the best of both Japanese and American production and labor relations systems. They now believe that many of the differences are superficial (e.g., everyone wears the same uniform and eats in the same cafeteria and no one has assigned parking spaces). Other differences, such as team-oriented production, are considered a mixed blessing. In addition, many workers contend that the Japanese managers are convinced that Japanese production and personnel practices are superior across the board to American approaches and are trying to impose these practices on the American work force. It is likely that Americans will continue to resist elements of the Japanese approach. Workers have come to accept, however, that, as in any automobile plant, decisions are made with an eye on the bottom line. It appears that most workers will accept this reality but will aggressively assert their interests, as they would in traditional U.S.-owned automotive firms. Others who cannot accept the conditions will join the procession of workers who have already left.

The Safety and Health System at Jidosha USA

Senior Jidosha management in Japan, including the CEO, with whom I met to discuss my research, repeatedly stressed the importance of worker safety and health at Jidosha USA. This is not unusual for managers, especially when speaking to a researcher concerned with occupational safety and health. His assertions were substantiated, however, by concrete corporate actions. Furthermore, Jidosha USA safety and health staff as well as outside observers, such as state-plan Occupational Safety and Health Administration officials, regard the commitment to safety and health by top management as one of the most significant differences between Jidosha and U.S.-owned automotive firms. Jidosha USA safety and health professionals who have worked in other U.S. auto firms characterize similar statements of support by their senior executives as "lip service." According to the safety and health director, there is nothing fundamentally new or different about the Jidosha safety and health system; it is "nothing more than very solid, very sound, simple common sense, basic safety techniques." According to him, it is the extent to which these techniques are implemented, made possible because of top management's support, that sets Jidosha's safety and health system apart from others with which he is familiar.

Senior Jidosha management recognized that the adaptation of their safety and health system to American workers and managers would be a significant challenge. Planning and management of safety and health operations and training of workers were therefore begun far in advance of the opening of the plant. The director of the Safety and Health Department was one of the first managers hired and came on board more than eighteen months before the official start of production.

Occupational safety and health activities at Jidosha USA are important for several reasons. The corporate philosophy and the system of manufacturing dictate that safety and health be viewed as intrinsic parts of the production process. As noted earlier, use of the Japanese system of flexible manufacturing, in which there are highly interdependent work stations and workers, small inventories, few redundant workers, and harmonious production, means that accidents seriously disrupt the production system. This system thus requires that safety and health receive the same degree of attention American manufacturers normally focus on productivity and quality. At Jidosha USA, safety and health are considered inseparable from these other aspects of production.

Safety and health activities at Jidosha USA are important for another reason: They are a way management can symbolically express its concern for workers. Relative to Japan, the United States has much more controversial labor-management relations regarding work in general and safety and health in particular. Thus, in that management's concern for safety and health has a bearing on the success of company efforts to develop more cooperative labor-management relations, it is very important that workers have favorable views about the safety and health program. It is also important that the general public see the company as a good corporate citizen,[3] especially in light of current difficulties in economic and trade relations between the United States and Japan and the sensitivity about Japanese investments in the United States. The promotion of worker safety and health can be crucial in creating cooperative labor-management relations and a good corporate image. In contrast, significant safety and health problems could have disastrous ramifications.

Organizational Structure and Staffing

The Safety and Health Department at Jidosha USA is divided into four parts: safety, industrial hygiene, security, and health services. Most of my attention will be focused on the first two components. In addition to the director of the Safety and Health Department and his clerical assistant, there are five safety specialists, one industrial hygiene specialist, and one Japanese adviser. Also reporting to the safety and health director are a health services coordinator, who oversees providers of medical and fitness services (including an elaborate fitness center on the grounds); an employee assistance program representative, who works with the health services coordinator; and three security supervisors, who oversee security service subcontractors. (The medical and fitness services were subsequently made a separate Health and Fitness Department.)

A UAW safety and health representative works very closely with the safety and health staff and is a working member of the Safety and Health Department. He reports directly to the safety and health director and is paid by Jidosha, although he estimates that 80 to 90 percent of his time is spent on union (as distinct from company) safety and health concerns. He is appointed by UAW headquarters but reports to the local elected union leader. His appointment, rather than election, to this position is consistent with practices in the rest of the industry. It is intended to protect the representative from union politics (see Wokutch 1990a).

In contrast to Jidosha in Japan, Jidosha USA has an employee assistance program. The program provides, by way of referrals to specialists outside the company, confidential assistance to wage and salary employees experiencing drug- or alcohol-related problems or other personal difficulties. (Job applicants are tested for drugs and alcohol, but there are no random or periodic tests. It is felt that such testing would alienate workers and be counterproductive. Tests are administered only when there is suspicion of drug or alcohol use based on a supervisor's observations.) To ensure additional confidentiality and minimize the likelihood of negative effects on anyone using the program, there are two referral coordinators. One is a UAW member, and the other is a member of management.

The safety and health director and most of the staff of his department had professional experience in the safety and health field before coming to Jidosha, and about half worked in the auto industry. Unlike the Jidosha safety and health staff in Japan, most also had academic training in the industrial safety and hygiene field before being hired. The Japanese Jidosha adviser had extensive safety and health experience with Jidosha and had successfully directed the implementation of the safety and health program at the new Jidosha plant in Japan. Upon his return to Japan, he was promoted to assistant manager of the Safety and Work Environment Division.

Safety and health activities during construction of the Jidosha USA facility were overseen by the manager of the Safety and Work Environment Division of Jidosha Japan, the second-ranking safety and health official. He was dispatched to the United States in large part because there were several serious injuries, including one fatality, during construction. His superiors were apparently pleased with his performance, for after he returned to Japan, he was named general supervisor of the Safety and Work Environment Division.

Core members of the Safety and Health Department at the U.S. plant were hired one year before the official start of operations. These included the director; three safety specialists, including the UAW safety and health representative; and the industrial hygienist.

The Safety and Health Department is in the same large work area as the other components of the Personnel Department and the Labor Relations Department. Small partitions are all that separate "offices," and managers had to fight to get them. This design facilitates interaction and group decision making, but there is much less privacy than in the typical U.S. office. The arrangement has been a source of irritation and for some has meant a loss of status. In addition, people have to be more aware of how their activities affect others, for example, how loudly they talk on the phone. (The safety and health director is notorious in this regard. He is routinely and good-naturedly harassed by his colleagues in Safety and Health, Personnel, and Labor Relations.) Nonetheless, Jidosha USA staff have far more space and privacy than do their counterparts in Japan.

In addition to safety and health staff, all line managers at Jidosha USA are considered part of the safety and health system. Many U.S. firms assert that "safety is everyone's business." Jidosha USA takes this slogan seriously. Specific safety and health duties are delineated for managers ranging from first-line supervisors to the manager in charge of manufacturing. I was provided a draft of a five-page oversized diagram (about sixteen inches by twelve inches) entitled "Safety and Work Environment Control Duties of Managers in Jidosha Manufacturing Divisions," prepared by the Safety and Work Environment Division of Jidosha in Japan. Jidosha trainers from Japan explained the diagram to Jidosha USA staff during their training. It details specific duties of the unit leader, section supervisor, department manager, division manager, and manufacturing manager on the following dimensions:

basic duties;
establishment and promotion of safety and work environment control;
drafting and promotion of safety and work environment control policy and program;
calling safety and work environment committee meeting and conference;
decision of safety and work environment budget;
carrying out previous evaluation and promotion of maintenance of newly established or changed machinery, etc.;
carrying out legal inspection program;
carrying out study of production and work methods taking into account safety and work environment;
preparation of work standards concerning safety and work environment;
thorough observance of laws, regulations, etc.;
preparation of legally qualified persons;
carrying out safety and work environment education and training program;
promotion of health control;
studying and carrying out proper placement;
carrying out safety and work environment patrol;
control over dangerous matters and harmful matters;
measurement, evaluation and improvement of work environment;
promotion of wearing safety and work environment protectors;
drafting and carrying out safety and work environment events;
record-keeping on safety and work environment;
investigation of causes for accidents, and to study and carry out measures to prevent recurrence.

Responsibilities of production workers were delineated elsewhere. Jidosha USA safety and health staff were responsible for ensuring that the rest of the U.S. managers and workers understood and carried out their responsibilities, thereby permitting the safety and health staff to act as consultants rather than as "firefighters." As we shall see, however, as production began, many managers and workers slipped into the routine common in U.S. plants of relying on safety and health staff. Further, there were some significant personnel changes after the time period on which this book focuses. Particularly noteworthy was the pressured resignation of the safety and health director. This came about, at least in part, because upper management was unhappy with safety and health conditions and the degree of labor-management conflict over them.

Goals and Objectives

According to the safety and health director, he was told when he joined Jidosha USA that his task was to create the best safety and health program, bar none, in the United States. He was requested to duplicate, within constraints imposed by legal and cultural considerations, the safety and health system of Jidosha in Japan.

This explicit attempt to model the safety and health program at the U.S. plant on the one in Japan is unusual. According to sources at other Japanese auto transplants, only one other transplant intentionally modeled its safety and health activities on those of its Japanese parent. The safety and health directors of the other transplants claimed they were given more discretion and less explicit direction in establishing their programs. Nonetheless, most safety and health programs ended up similar to those of their parents.

As in Japan, workplace safety and health considerations at Jidosha USA are firmly entrenched in product and process planning and design, most of which takes place in Japan. Jidosha USA staff merely provide feedback to the corporate engineering staff. The integration of process planning and design is more systematic in the United States, however. There is a formal thirteen-step planning and evaluation procedure for new machinery and processes before they are integrated into production. Occupational safety and health issues are formally considered at several of these stages and the safety and health staff are heavily involved. Hazard-prediction procedures (KYT activities) are used during this process to identify the most likely and most serious problems, and countermeasures are then designed to address the hazards. The intent was to follow these procedures for all new machinery, and at the start of production, safety and health, engineering, and production personnel did so for virtually all major equipment. As the pace of production picked up, however, many modifications to production processes were made without following all of these steps.

Initially, managers at Jidosha in Japan spoke of the need to develop a safety and health system at Jidosha USA that would combine the best features of the Japanese and American systems. By the time production began, however, the Japanese managers (and many Americans as well) felt that the Japanese system was clearly superior, and that except for those required by law, relatively few features of the U.S. system should be used. This was the same pattern as emerged in attempts to combine other Japanese and American personnel and production practices.

The stated intention of the Jidosha USA safety and health director is to use Jidosha's safety and health system in its entirety, including its goals and objectives, with only minor changes. As we shall see, more significant than the differences in goals or activities are the differences in the way these goals and activities are implemented. Moreover, as production has proceeded and problems have developed, managers have been forced to adopt many safety and health measures more in keeping with the U.S. system than the Japanese.

The safety and health director contended early in my research that there were no budgetary constraints on his pursuit of safety and health goals. When asked what he would do differently with a 50 percent increase in his budget, his response was "nothing," because he sets the budget according to the needs of the organization. This was clearly an exaggeration of his authority, of the importance attached to safety and health, or both. The problems with cumulative trauma disorders that developed in the first two years of operation and that could have been more quickly addressed by redesigning work processes belie his contention.

In addition, one member of the safety and health staff in complaining about the workload contended that the department was "understaffed." The director's response to this comment was that this individual did not understand the Jidosha safety and health system. According to him, the staff needed to get line managers and workers to fulfill their responsibilities regarding safety and health so that his staff could act more as consultants and trainers rather than firefighters. The problem, he said, was not the level of staffing.

In keeping with the Jidosha safety and health system, Jidosha USA plans to adopt a version of the zero-accident campaign. The near-miss accident reporting system is already in place. The rationale for this system is that a near miss indicates an underlying problem or defect that must be corrected. When I said that a near miss was "luck" (i.e., that it did not result in an injury), the safety and health staff objected. In their view, luck has nothing to do with it. The question is whether the worker avoided a point of impact.

At this stage of development, the zero-accident campaign is more of a philosophy than a realistic goal. Realistic injury and illness goals are set for the company and for individual departments, and 20 to 25 percent of the bonus available to managers is based on whether their departments' injury and illness rates are in line with these goals.

In calculating rates, officially recordable injuries and illnesses and what are referred to as "Jidosha emphasis injuries" are distinguished from each other. The latter include serious injuries such as lacerations requiring more than first aid, fractures, loss of consciousness, and any lost-time injuries and illnesses. Other less serious injuries and illnesses that require medical attention are recordable according to state-plan OSHA guidelines. This parallel recordkeeping system was devised by the safety and health director. Its stated purpose was to focus managers' and workers' attention on eliminating serious injuries and illnesses. He believed it was necessary because of what he described as the overrecording of injuries and illnesses for SOSHA purposes in the wake of the fines OSHA imposed on auto firms for underrecording. Perhaps an underlying motive was to distinguish cumulative trauma disorders, which were fast becoming a problem, from Jidosha emphasis injuries (CTDs would not be counted in this category unless they resulted in days away from work). The Safety and Health Department could thus show evidence of addressing traditional disabilities while shifting attention away from CTDs.

The use of numerical injury and illness rate-reduction goals in evaluations of managers was a sensitive issue with management and was being reconsidered when I visited in light of OSHA's fines of the Big Three automakers for recordkeeping violations. One of the principal reasons the companies altered the numbers of injuries and illnesses they reported was to escape OSHA inspections, which are conducted most frequently at companies and establishments with injury and illness rates above industry averages. Since these fines were imposed, the industry has tended to focus less on numerical goals for fear of incurring recordkeeping violations or arousing OSHA's suspicions of such. Thus, at Jidosha USA numerical injury and illness rate-reduction goals are now augmented by process-oriented safety and health goals.

In setting goals and in planning activities, Safety and Health Department staff, as well as workers and managers throughout the plant, have experienced some frustrations with the Jidosha planning and decision-making process. The decision-making system, as in most Japanese firms, requires participation of all members of the group. Naturally, this slows down decision making and irritates safety and health staff members who are accustomed to making quick decisions, especially those who have worked at traditional U.S.-owned automotive plants. One safety and health staff member noted, however, that, although decisions at his former employer were made more quickly, everyone fought over questions of implementation. He said that at Jidosha there is less bickering at the implementation stage.

Another frustration of some staffers was that discretion in planning is limited. They complained that their decision making is constrained by the "Jidosha plan" imposed from Japan. Activities and expenditures that conflict with this plan are considered out of the question. This apparently resulted in a lag in recognizing and responding to problems such as CTDs. In the words of one disgruntled staffer, "If it's not in the Jidosha plan, it doesn't exist."

Prestart-up Activities

Several safety and health activities had already been begun before the start of operations. Certainly the most important such activity was training. The training program again illustrates how safety and health was integrated into the overall scheme of operations.

Training. Like other Japanese auto companies that have located production in the United States, Jidosha was careful to ensure that the American work force was adequately trained in the exacting standards of Japanese production. There were many months of training and runs of the production line before the start of operations.

Essential to the overall training program was training for team leaders, supervisors, and managers in Japan, where in addition to skills training, employees were inculcated with the Jidosha culture of work. Groups of sixty to seventy employees were trained together, but, in that the staff consisted of approximately thirty trainers, fifteen translators, and five program administrators, each trainee had plenty of individual attention.

Workers were first given two weeks of instruction in the United States, consisting of a general introduction to Jidosha's philosophy and a description of its technology and engineering systems. Throughout this process, efforts were made to promote "friendship and morale" among the workers. This initial training was followed by a four-week training session in Japan that consisted of three components: (1) demonstration that the concept of operation taught in the United States actually works at Jidosha; (2) experience with the production system; and (3) enhancement of friendship among American and Japanese Jidosha workers, especially Japanese Jidosha trainers.

Aspects of safety and health were incorporated throughout the training in Japan but were especially emphasized in several sessions. First, workers received a two-hour lecture on general work safety and health issues. Second, they received a half-day of safety and health training at their workshops. And, third, they received individualized training at their work sites. Training was reinforced during the start-up of operations in the United States.

In addition to team leaders, supervisors, and managers, all members of the Safety and Health Department were sent to Japan for training.

Their training consisted of classroom instruction in the Jidosha safety and health system; training exercises (for example, fault tree analysis, in which trainees examine a mock-up of an accident to determine its cause); and tours of Jidosha facilities. A description of the overall philosophy of safety and health activities at Jidosha, as well as specific activities and the organizational structure, were explained in excruciating detail.

The UAW safety and health representative especially benefited since he had no experience or training in safety and health. Likewise, his participation yielded tremendous benefits in terms of team building. The safety and health staff obviously developed a sense of camaraderie.

I observed portions of the training program for team leaders, supervisors, and managers firsthand and interviewed a group of trainees (selected by their Japanese trainers) in the lobby of the hotel where they and I were staying. The six trainees, all of whom would be team leaders or unit leaders when production began, had been carefully chosen. They all spoke about Jidosha with the enthusiasm of recent converts. And including a woman and a black in the group was not lost on either the trainees (one of whom made a comment about it) or me. I elicited other opinions about training in informal conversations with trainees whom I encountered at the hotel. I also met informally with four members of the Safety and Health Department.

Although it requires a digression, the circumstances surrounding the arrangement of these meetings provide some insight into the challenge of conducting research in Japan. Initially, Jidosha safety and health staff in Japan resisted my request to visit the training facility. The response was always something to the effect that "that is very difficult" or "their schedule is very busy." They grudgingly conceded to let me go to the facility only after I discussed the problem with the semiretired senior Jidosha official who had arranged my visit to the small subcontractor. He went over the head of the general supervisor of the Safety and Work Environment Division to a senior member of Jidosha management, who subsequently directed the general supervisor to arrange a visit to the facility.

My meeting with the safety and health trainees can only be attributed to luck. Jidosha safety and health staff were even more resistant to arranging a meeting with them than they were to allowing me to visit the training facility. Although I had repeatedly expressed an interest in meeting with the trainees, Jidosha safety and health staff apparently hoped to hide their presence from me—their stay in Japan would overlap with mine for only about two weeks—so as to avoid the embarrassment of having to deny my request. This became clear when a safety and health staff member let it slip that the trainees were coming to town near the end of my stay. Although I speak very little Japanese, it was clear that his supervisor reprimanded him for this slip and that he reprimanded him again more strongly when I asked to meet with the trainees. Again, the response was that they were too busy and that it would be difficult to arrange. This is where the element of luck came into play. I had given small "Virginia Tech" pins I had brought from my university as tokens of appreciation to all the Jidosha staff in Japan with whom I had worked. A staff member was wearing one of these pins when a Jidosha USA trainee inquired where he had gotten it. As luck would have it, I had interviewed this trainee on several occasions in his previous job at a Big Three auto plant. When he learned I was in town, he insisted on getting my phone number, and the Japanese, always the gracious hosts, could not refuse. When he called, I arranged to meet him and the rest of the safety and health trainees for dinner. Dinner and drinks at a Japanese country-western bar provided a casual atmosphere in which the trainees could talk about their experiences in the training program and speculate on how well the Jidosha safety and health system would work in the U.S. plant.

The proffered explanation for why Jidosha staff in Japan were so reluctant for me to meet with the trainees lacks credibility, for, although their schedules were busy, there was certainly slack time. Two more plausible explanations are that there were aspects of the training that the staff did not want me to learn about or matters the staff did not want me to discuss with the trainees, especially those from safety and health. While these explanations seem plausible, the exact reason for the secrecy remains a mystery.

Once the visit to the training facility was arranged, I learned that both a translator and a senior safety and health staff member would accompany me. The stated reason was to "assist" me, but as it became clear, they were also there to keep an eye on me. In addition, a tight schedule was set up, apparently to keep me busy. It was only when I "escaped" from my escorts that I was able to have some chance contacts, which proved to be far more informative than the formal interviews.

After all the difficulty involved in arranging to see them, I was surprised that the trainees were enthusiastic about Jidosha, for by this time I suspected that management was hiding a significant problem from me. They were impressed with the Jidosha production system and the training program and pleased with how welcome top management, trainers, other Jidosha employees, and even local Japanese unaffiliated with Jidosha had made them feel. They indicated that they had the feeling that they were part of a team. They especially valued this feeling since potential union members (team leaders) and salaried workers (unit leaders and other managers) were being trained together. They firmly believed that the sense of being part of one team or family would carry over to operations in the United States.

The trainees also welcomed the detailed instructions they received. They were given worksheets with instructions on how to perform every step of their jobs, including, for example, where to stand and which hand to use. The worksheets also contained warnings about where the hazards were on each job and how to avoid them. One trainee noted that he had never seen anything like these worksheets in the six years he had worked in three different departments of General Motors' Cadillac Division.

Trainees were required to familiarize themselves thoroughly with the worksheets before beginning work. As one trainee said,

> You were made to look at it ten minutes at least before you even look at anything else. Then when you get out to the line they make you look for a half hour, if not longer, and they point out every move, the little flicks of the wrist or the way the body is positioned so the weight's distributed correctly, or the hazard points, like you know, the reason the guy's holding the deck light up is because it won't fall down on him. They point all that out to you. Lots of times I found myself saying, "Yeah, well, when can I work? When can I work?" They say, "No, slow down, learn your job."

Trainers indicated that they expected workers to take between two and four weeks to learn assembly line jobs and that greater proficiency would come later. One trainee noted that workers at the U.S. auto company where he formerly worked had had three days to learn their jobs and were fired if they could not learn them in this time.

The trainees were very impressed with the safety and health system at Jidosha and with the emphasis on safety and health in the training. The workers believed that their safety and health was the first priority of the trainers, even more important than production and quality. (In keeping with Japanese tradition, the trainer, not the injured trainee, would be reprimanded if an injury occurred.) Workers felt that this concern for safety and health was indicative of management's concern for them as people, which they anticipated would affect other areas of labor-management relations.

The trainees also felt that the work was conducive to promoting safety and health and that the jobs were "fun." The pace of work during training was described as steady and fast enough to "work up a sweat" but not so fast as to leave them physically exhausted at the end of the day. They noted, however, that the speed was well below the projected rate for the line in the United States. Nevertheless, they were confident that the speed would be set with American workers in mind, in other words, slower than in Japan. One trainee described his experience this way:

> I was put on a job that I was doing, and I was thinking, I said, "how in the world does this [Japanese] guy do this job all day?" I was thinking that to myself. But that's a Japan cultural thing. I think that when the jobs are put in and time studied and engineered into [the U.S. plant] they're taking consideration of an American worker. I don't think it can be, I don't think those jobs for American workers can be done as efficiently so to speak so far as time is concerned in America as they are here because of the difference in the people. These people when they go to work they're there to do their job. They don't hardly talk. The little guy that I was working with, I had to almost break his neck to make him say "hi" to me because he's so busy concentrating on his job.

The work pace differed from the jobs some of these workers had held with U.S. auto manufacturers. In those plants workers could have a drink, smoke, or even read a book if they worked fast enough to build up a buffer between their work station and that of the next job. At times this would affect their concentration or cause them to rush and get injured. (In talking about this problem, a state-plan OSHA safety and health official described an accident in which a worker who had built up a buffer went to sleep. He subsequently broke his wrist when he fell off his chair and the pallet on which it was sitting.) Moreover, the trainees were taught and believed (or at least said they did) that 100 percent of a worker's time on the job belongs to the company and that a person should be contributing to production at all times during working hours. Thus, like their Japanese counterparts, they were instructed to clean up their work area, sharpen tools, or do preventive maintenance during line stoppages.

Trainees provided a different and conflicting view about the work pace over drinks in a more informal setting, a hotel suite set aside as a lounge for the trainees. Several expressed the view that American workers would not (or could not) tolerate as fast a work pace as Japanese workers do. They were in awe of the speed and concentration of the Japanese workers with whom they had come in contact. Whether American workers can and will work at this pace is still an open question. The fast pace is one of the most common complaints among workers at Jidosha USA and was critical to the success of a dissident slate of UAW candidates in local elections. Moreover, two years into operation it would be extremely difficult to find any Jidosha USA production workers who would characterize their jobs as "fun."

The trainees were generally impressed with the way safety and health were viewed as the responsibility of line managers and individual workers rather than of safety and health professionals. Most felt that Jidosha USA managers and workers would be able to adapt to Jidosha's view since they would understand the situation from the start. They conjectured that it would have been harder if they had had to change their behavior in an existing enterprise. A few trainees questioned whether production managers would accept responsibility for safety and health in their work areas, rather than relying on safety and health staff, but others were confident the safety and health staff would be successful in helping managers accept their responsibilities, remarking that they were "pretty sharp."

Two years into production, however, efforts to get line managers and workers to assume these responsibilities have been only a modest success. Safety and health staff complain that they continue to receive requests to correct hazards that individual workers or line managers should be able to take care of themselves. Further, a disturbing pattern has arisen whereby some unit and team leaders compromise safety and health considerations when they conflict with production concerns.

Despite the favorable opinions of the training program overall, there were complaints, as one might expect. Most of the workers admitted to being homesick and doubted that the training program needed to be so long. They thought the in-class lecture material could have been compressed or presented at the U.S. plant, and some thought that the material was "too philosophical." Others complained that activities were too planned and structured for them. Some expressed an interest in having more free time to sightsee, shop, or do other activities. Apparently they were not completely comfortable with the strong embrace of the Jidosha family. Trainees were also impatient with the group decision making required by the *kaizen* process before ideas could be implemented. On the day I visited, one trainee had stormed out of the room because he had become frustrated with the process in his work group. Needless to say, this show of temper did not go over well with the trainers. In fact, both trainers and trainees readily conceded that Americans do not work as well in groups as the Japanese do. Moreover, since the trainees were expected to teach *kaizen* activities to production workers in the United States, further problems were anticipated. This difficulty working in groups combined with impatience is at the heart of several problems American workers have had in adapting to the Jidosha style.

Several workers complained that their trainers criticized them for what seemed to be minor or petty offenses, such as being a few minutes late for a bus or slouching while listening to a lecture. In both cases the workers resented being shamed and, in their view, treated like children.

These difficulties obviously reflect intercultural differences between Americans and Japanese that will present continuing problems at Jidosha USA. They also reflect different views regarding authority and obedience. Japanese managers are accustomed to full and unquestioned compliance from subordinates. To them, the American trainees seemed unruly. To the American trainees, the training was much like military boot camp. In that the training program was designed as much to transmit Jidosha's corporate culture as to provide training in technical aspects of production, continuing problems at Jidosha USA indicate that the American workers have not fully accepted this culture.

One incident a trainee related to me further demonstrates the cultural differences between Japanese and U.S. styles of worker protection. He described how he had observed what he felt was a serious hazard at the Japanese plant. A large barrel of extremely hazardous chemicals called isocyanates sat unprotected in the center of the factory floor with no more than a "danger" sign on it. When the trainee pointed out the hazard to the trainers, they responded that workers would see the sign and know to avoid it. The trainee could only imagine the disastrous consequences that would result if the barrel were accidentally hit by a forklift.

This incident illustrates the greater reliance in Japan on a behavioral approach to safety and health management. In the United States, there is a tendency to use engineering solutions in an effort to make certain work conditions are "absolutely safe." Thus, in the U.S. Jidosha plant, isocyanates are stored in a concrete structure.

Like Japanese workers, the American trainees were unwilling to report minor injuries. One woman said she did not want to be known as the first one to have an accident, especially because the trainers emphasized safety so much. As a result, she concealed some minor cuts. Other trainees conceded that they had also hidden minor cuts or sprains.

Following the training program in Japan, all Jidosha USA employees received training in the U.S. facility. Training consisted of a three-week program for everyone in which two full days were devoted to safety and health. Workers then received individualized training in the specific functions they would be performing. Workers in some functions including robot operations, welding, and hoist operations needed to be certified before they could proceed further. Programmed worksheets describing precisely how jobs were to be performed and incorporating safety and health guidelines were provided at each work site, as well as further individualized and group training.

Subsequent hires did not go through nearly as extensive a training program. In many cases, they were needed on the production line immediately and their training was "quick and dirty." Ironically, some early Jidosha hires had criticized their former employers for this practice. Union leaders claim part-time workers receive particularly inadequate training. Management claims that their training is adequate in the specific job functions they are to perform.

Current Operations

As it did for the goals and objectives, the Jidosha USA safety and health staff asserted that it intended to adopt the entire range of Jidosha safety and health activities. (They subsequently conceded lightheartedly that the safety and health prayer with government regulators probably would not go over well in the United States.) Most of the features of Jidosha's safety and health operations in Japan were already being used at Jidosha USA two years into operation. These activities included training, safety and health committee meetings, inspections, the use of safety signs, hazard-prediction and hazard-prevention activities, and accident reports and investigations.

Training. Follow-up safety and health training is proceeding slowly. The safety and health staff would like to provide training in aspects of hazard prediction and prevention that have not been taught, but because of the pace of production, finding the time is a problem. Further, as noted earlier, new hires are frequently rushed through training. In contrast, in Japan, only a few new workers are usually hired at a time, and they can usually be replaced if they need to go for training.

Training difficulties at Jidosha USA are further exacerbated by the increase in the size of the work force and the turnover, especially among management. New managers have not been fully trained themselves and consequently do not place the same emphasis on safety and health as managers who have had the full complement of training. In addition, they are less willing to free up their subordinates for training.

Committees. As with Jidosha in Japan, safety and health committees operate at all levels of the organization. At the highest level there is a safety and health committee chaired by the executive vice-president of administration, consisting of the executive vice-president of manufacturing, division managers, department managers, the safety and health director, the local UAW shop chairman, and the UAW safety and health representative. Management is obviously much more heavily represented in this committee than in the comparable committee in Japan. The committee meets every month to consider safety and health issues in the plant. If there was a recordable injury (one requiring medical intervention other than first aid), the responsible manager is required to explain what happened and what was done to correct the situation.

Each division and each department also have labor-management safety and health committees. At the department level (e.g., Paint Department, Materials Handling Department, and so on), safety and health activities are directed by a management safety and health coordinator and a UAW district committee member.

At lower levels of the organization (areas, units, and teams), the requirements regarding safety and health committees and meetings are less uniform. Some department managers require area and unit meetings, other do not. All units are required to give thirty-minute safety and health talks each month.

At the team level, safety and health meetings are held on a voluntary basis. Teams that elect to hold such meetings usually do so in the morning on their own time. There is some pressure between and within teams to engage in this activity but not nearly as much as in Japan. Some teams also have a safety and health coordinator, but, again, one is not required.

Inspections. As in Japan, inspections are conducted both formally and informally. Each department manager assigns someone to conduct an inspection as regularly as he or she sees fit. The Safety and Health Department monitors these inspections to ensure that departments are fulfilling their obligations, and safety and health staffers also inspect on a regular basis. Unlike traditional U.S. plants, however, safety and health staff members at Jidosha USA do not have the authority to "red-tag" (shut down) a piece of machinery for safety or health reasons unless there is imminent danger. They can *recommend* shutting down a machine, but the production manager decides whether to do so. Compared with their counterparts in Big Three plants, production managers at Jidosha have greater authority to implement safety and health goals and safety and health professionals have less authority. The red tagging of machines by safety and health staff at Big Three plants occasionally leads to acrimonious disputes with line managers who see such an action as infringing on their authority (Wokutch 1990a). According to the espoused safety and health philosophy, such disputes should not occur here because line managers have supposedly internalized safety and health goals and values. In reality, they do not occur at Jidosha because the safety and health staff do not have the authority to confront line managers the way they do at Big Three plants.

Production workers are encouraged to report any problems and to file near-miss accident reports when a mishap occurs that could have caused an injury. Workers are quick to register complaints. Managers consider most of their complaints to be about comfort, but workers regard them as legitimate safety and health concerns. Notable are complaints about the high temperature during the summer and about odors in production areas. The heat ranks right behind the pace of the work as a source of discontent.

Signs. There are a great many signs, including safety signs, throughout the plant. These provide both motivational and informational messages on a wide range of work-related issues. By and large, they are similar to the signs in Jidosha's plants in Japan, but some address hazards unique to the plant in the United States. For example, some of the women and men have long hair that could get caught in a machine, so there are signs exhorting them to cut their hair short or tie it up. One sign graphically makes this point; it shows a drawing of a woman with her hair ripped off. There are, of course, no women production workers or, for that matter, men with long hair, in the Japanese plants.

Hazard-Prediction Activities. Workers at Jidosha USA have been trained in several of the hazard-prediction and hazard-prevention (KYT) activities used in Japan. The acceptance of these activities has varied, but none is accepted as much as in Japan. Workers have been trained in the use of the morning safety check, for example, but only rarely is it used. The formality and ritualism of this activity are difficult for American workers to embrace. One safety and health professional described it as "not macho."

Initially, there was widespread acceptance of and reliance on the Jidosha-developed SKY (immediate hazard-prediction) cards for use when entering machine work areas. The Jidosha USA safety and health staff even came up with an innovative adaptation for American workers. On the wall in the area where the cards are stored, each worker posts a picture of a loved one. The idea is to remind workers to be careful not just for themselves but for others. Such a mixing of work and home life would be out of character for Japanese workers. Nonetheless, Jidosha USA had to institute lockout procedures like those used in the rest of the industry for use when a worker enters a machine work area. The change came about through an agreement between labor and management, and SKY cards are now used in combination with traditional lockout procedures. As one safety and health staff member said, they were worried some "cowboy" would start the machine while someone was in the work area.

Jidosha USA safety and health staffers initially hoped to have the full range of KYT activities used by Jidosha in Japan. Gaining full acceptance of the activities is difficult, however, and it is unlikely that they will ever all be used.

Accident Investigations. Jidosha USA accident investigations follow the same general procedures and guidelines as at Jidosha in Japan. Any serious accident is reported back to Japan. Contrary to intentions, however, there is neither a systematic corporatewide reporting procedure nor common definitions. Jidosha USA records injuries and illnesses according to state-plan OSHA guidelines, and Jidosha in Japan records them according to the various definitions described in the preceding chapter. Injury and illness statistics for Jidosha USA are reported to corporate headquarters as well as to state-plan OSHA offices.

Despite extensive efforts to understand the differences in definitions in the United States and Japan, there is still considerable confusion at Jidosha USA. The director of safety and health noted that the definitions have been an ongoing matter of discussion since he started at Jidosha, and as far as he was concerned, he was no closer to resolution of the matter. He complained that he got different answers to his questions depending on whom he asked. Undoubtedly the poor English-language skills of the safety and health advisers whom he asked contributed to the confusion. Nonetheless, it is generally agreed that the Japanese and U.S. definitions are very similar.

Unlike the practice in Japan, Jidosha USA's injury and illness experience is not compared with that of other firms in the industry. There is currently no mechanism for comparisons like the one provided by the safety and health subcommittee of the Japanese Automobile Manufacturers' Association. Before he resigned, the safety and health director at Jidosha USA was in communication with safety and health managers at other Japanese-owned automotive plants in the United States in an attempt to form a similar association. Should the association ever get started, data on work injuries and illnesses could be shared and compared under its aegis.

Relations with Regulators

One of the most interesting aspects of the safety and health system at Jidosha USA is the cooperative relationship it has developed with the state-plan OSHA (SOSHA) that has jurisdiction over the plant. Efforts to develop this relationship began well before the start of operations. In an unprecedented move, Jidosha invited several SOSHA officials to inspect Jidosha facilities in Japan as well as machinery under construction by Japanese subcontractors of Jidosha scheduled to go into the U.S. plant. The Japanese safety and health adviser, who may well have been instructed to initiate the process by his superiors in Japan, suggested the idea to the American safety and health director. The director and his American staff, who had been so impressed with the safety and health conditions in Japan, endorsed the idea wholeheartedly. Two contingents of SOSHA officials subsequently visited Jidosha and Jidosha subcontractor facilities in Japan.

The SOSHA officials who made the trip noted that other firms occasionally request advice during preproduction regarding SOSHA compliance. This was the first time, however, that officials had been invited to visit an existing plant and to travel out of the country. SOSHA officials enthusiastically accepted the invitation. The trip was a fascinating experience for the officials, and the state saw it as an inexpensive way to accommodate Jidosha. Everyone was delighted with the arrangement.

The stated reason for the trip was to get advice on whether or not the machinery and facilities planned for the Jidosha USA plant would meet SOSHA standards and to get advice on compliance where it would not. The underlying aims were just as important, however. By accompanying the SOSHA officials around Japan and treating them with typically gracious Japanese hospitality (what we might term "wining and dining"), Jidosha officials (both American and Japanese) were able to develop a good rapport and sense of fellowship with their SOSHA guests. Clearly, the Japanese hoped to co-opt their guests as well, and to a large extent they were successful. The cooperative relationship that developed contrasts sharply with the adversarial one with SOSHA that safety and health experts at other auto companies describe but is consistent with regulatory relations in Japan. In most cases, the company's safety and health standards met or exceeded those set by SOSHA. Jidosha USA managers can therefore pursue their safety and health activities without looking over their shoulders for SOSHA officials. Further, Jidosha USA officials feel comfortable inviting SOSHA officers into the plant in an unofficial capacity to seek their advice. Significantly, SOSHA officials have asked company safety and health managers for advice on several issues for which Jidosha practices are perceived to be superior to SOSHA requirements. They have also involved Jidosha USA staff in at least one SOSHA-sponsored training program for safety and health professionals at another firm. Again, this behavior is more common in Japan than in the United States.

The first official SOSHA inspection was conducted eleven months into production. It was done at the request of an unidentified employee who contacted SOSHA to complain about gas fumes and leaks in the area where new cars are filled with gasoline before they are test driven. Apparently a leaky nozzle had caused the problem, but according to the safety and health director, it had been corrected well before the inspection was conducted and, he suspected, before it was requested. Inspectors gave the plant a clean bill of health. Nevertheless, the request indicated a breakdown in communication (i.e., the worker did not know that the situation had been corrected), a lack of trust that the problem could be solved internally, and probably some disgruntlement on the part of the worker who filed the complaint.

Later, a more comprehensive "wall-to-wall" inspection was conducted. It resulted in forty-five citations, three of which were classified as serious. Safety and health staff refused to provide me with details, but I obtained a report of the inspection from SOSHA under the state's Freedom of Information Act. A wide variety of conditions were cited, including improper storage of flammable materials, clutter blocking an exit, improper guarding of electrical wires, and inadequacies in the guarding of machinery. The serious violations were for (1) a nonfunctioning interlock system, which was supposed to shut down an industrial robot if someone entered the robot's work area (workers bypassed the system); (2) failure to have wheels of a truck blocked while it was being boarded by a forklift during unloading operations; and (3) employees overriding lockout procedures on several machines and entering the machine work areas while they were operating. Many of the violations occurred because workers were apparently undermining established safety procedures or systems in an attempt to speed up work and improve productivity.

Management was concerned about the citations but contended that forty-five was not an unusually high number for a facility of Jidosha's size. The results compared favorably with those of recent OSHA inspections at Big Three plants. Of greater concern were the three serious violations, which carried penalties totaling $1,330. If these violations were ever cited again, the plant would be charged with willful violations, which carry substantially higher penalties.

Management claimed that the "serious" designation was extreme and appealed to SOSHA. Jidosha USA conceded that, with respect to items one and three, the company was in technical violation of SOSHA guidelines but that, in light of its overall safety and health program, these transgressions did not merit being classified as serious. Management argued that the robot guarding and lockout systems were superior to those required by SOSHA, and even though workers were bypassing these systems, no one was in danger.

In the view of Jidosha USA management, the appeal hearing was conducted amicably and had favorable results. Violations one and three were downgraded from serious, and the total penalties were reduced to

$280. Management attributed its success to the good rapport it had with SOSHA officials and to their understanding of Jidosha's overall safety and health program.

Although management did not admit it, the results of the inspection, even after the results were downgraded, were undoubtedly disappointing. The results also raise serious questions about management's commitment to safety and health and its contention that safety and health procedures and practices at Jidosha USA are far superior to those in the rest of the U.S. auto industry.

Viewed in the most favorable light, the sabotaging of safety and health systems and workers' failure to follow procedures may have occurred because management failed to convince workers of the importance of safety and health.

A less favorable interpretation is that the violations accurately reflect management's emphasis on production at the expense of safety and health. Team leaders and unit leaders, who certainly should have noticed that workers were taking shortcuts, may have failed to stop them because safety and health are not nearly as important as managers claim when they argue that safety, quality, and productivity are mutually consistent goals. The Safety and Health Department certainly shares some of the blame. It is unclear whether staff members did not know of the unsafe practices or whether they knew of them and did not or could not stop them.

It is important to keep the results of the inspection in perspective. Relative to other auto firms OSHA has inspected in recent years, Jidosha USA did not fare badly.

Union Activities

The UAW's involvement in safety and health activities at Jidosha USA is distinctive from its involvement in safety and health at other U.S. auto plants. It also differs from Jidosha union activities in Japan. In particular, Jidosha USA is not a party to the 1973 Memorandum of Understanding on Safety and Health that the UAW signed with companies in the auto industry. Of course, Jidosha USA was not in existence at that time. No attempt was made, however, to negotiate blanket adoption of this agreement. Rather, most of the provisions of the agreement and subsequent amendments were negotiated separately and are now in the contract.

From both labor and management accounts, there was some contentiousness over safety and health issues in the contract negotiations that took place about one year into operation. Union demands pertained to revisions of the roles and responsibilities of the safety and health committees, lockout procedures, and the operations of a joint labor-management ergonomics committee. Members of the safety and health staff were surprised and disappointed that safety and health was an issue in negotiations since they felt that conditions were much better than in the rest of the industry. According to the safety and health director, management simply conceded to the union's demands. Management suggested that the UAW supply the language for the safety and health provisions of the contract and said it would sign it as long as the provisions were reasonable.

In addition to the full-time union safety and health representative at Jidosha USA, union input into safety and health is provided through eight district committee members (four for each shift). Each is responsible for representation of worker safety and health interests and of broader worker interests for two departments apiece.

To date, the union has not sponsored complementary safety and health programs such as safety committees, safety awards, and so on as the Jidosha union does. Overall, the U.S. union is much less heavily involved in promoting safety and health activities than the union is in Japan. This is in part because the structure of unions in Japan is different. In Japan, with its tradition of enterprise unionism, the Jidosha union is independent, whereas the union in the United States is a local affiliate of the huge UAW. On the one hand, because the local is not nearly as large as the Jidosha union and does not have all of its resources, even if the local wanted to, it could not engage in all the activities the Jidosha union sponsors. On the other hand, because the UAW is much larger than the Jidosha union, it has a far more diversified constituency and can concentrate only a small percentage of its attention and resources on Jidosha USA. The apparent desire of management to have the local act like a Japanese enterprise union in matters pertaining to safety and health is far from a realized goal. Moreover, the success of the New Directions movement in union elections suggests that the union may well resist proposals to be a partner as a matter of principle.

Cooperation and Conflict

Consideration of the cooperative and conflictual aspects of relationships at Jidosha USA vis-à-vis safety and health is intriguing because both are strongly represented here. Clearly, the relationships among labor, management, and government are in many ways cooperative. The close working relationship between the safety and health staff and the union safety and health representative provides the framework for cooperation between labor and management. Labor and management safety and health officials work very much like a team and are viewed as such by both workers and managers.

Especially during the early months of production when enthusiasm was high, workers and managers alike obviously believed the safety and health slogans at Jidosha USA, and most workers and managers still believe that safety and health are integral parts of the production system. This belief was emphasized during the training program and has been reinforced by the attention given to the near-miss accident reporting system. In the early stages of operation, workers expressed confidence, in fact overconfidence, in the ability of labor and management to work together constructively on safety and health matters as well as other mutual concerns. As one trainee said,

> One of the things Jidosha stresses is human dignity. That's one of the reasons why [unions] came about, because human dignity was not being adhered to. I've seen it many times here on the line, and I'm sure I'll see it many more times. It's safety first and human dignity first, and with that type of attitude, the union is going to be that much easier and more fun to work with.

Jidosha's safety and health committees provide an important vehicle for promoting cooperation between labor and management, and, because of group dynamics in this country, they may be of more benefit in the U.S. plant. Typically, U.S. workers need a role with some authority before they will pressure co-workers into conforming with safety and health norms. The committee structure increases the solidarity among workers while giving them the authority they need. Conformity with organizational norms pertaining to safety and health can be achieved far more easily in Japan without such a structure.

Despite the cooperative orientation to labor-management relations at Jidosha USA, there are clearly conflicts as well. These are evident in the results of the SOSHA inspection; in the continuing problems with injuries and illnesses, especially cumulative trauma disorders; in disputes over workers' compensation claims; and in workers' and managers' complaints.

The nature of the SOSHA violations indicate that safety and health concerns have been sacrificed to a certain extent in the pursuit of productivity, even though the company uses slogans such as "There can be no production without safety" and "Safety, quality and production are all part of the same operating equation." Such contradictions undoubtedly contribute to creating cynicism about management's commitment to safety and health.

Even more alarming is the high incidence of stress-related cumulative trauma disorders, including carpal tunnel syndrome, relative to the rest of the U.S. auto industry. CTDs result from the stress that occurs when specific motions are repeated, especially those involving lifting or pulling heavy weights, kneeling down, bending from the waist, or using vibrating machinery such as drills. Labor and management have cited various reasons for the problem at Jidosha USA: (1) the nature of the production system; (2) insufficient early attention to ergonomic modification of production machinery to suit the American work force; and (3) the inexperience of the workers.

Critics (Mirer 1989; Parker and Slaughter 1988) have argued that the Japanese production process in general and the Jidosha process in particular will inevitably result in a higher rate of CTDs than in traditional plants because of the emphasis on speed and on precisely defined and therefore repetitive motions. (In traditional U.S. plants, workers vary their motions throughout the day.) Moreover, the company pressures workers to use 100 percent of their time in production-related activities, leaving little time except for scheduled breaks when they are not performing work-related tasks. In defense of these practices, Jidosha USA officials contend that job rotation throughout the day minimizes the stress on any one part of a worker's body.

Jidosha USA officials generally concede that insufficient attention has been given to modifying Jidosha's production processes and machinery to fit the greater physical heterogeneity of the U.S. work force. In Japan, Jidosha production workers are all of the same sex and have similar body dimensions and strength. Machinery and production processes are designed for the typical Japanese production worker. A machine may have been designed for a five-foot, six-inch, 135-pound Japanese male, but at Jidosha USA a six-foot, three-inch, 220-pound male or a five-foot, three-inch, 115-pound female may be assigned to it. The joint labor-management ergonomics committee is viewed as an appropriate, albeit late, response to this problem. Both labor and management officials say the committee has been successful in getting ergonomic changes made to machinery and work processes, which should reduce CTD-related problems.

Another explanation for the high incidence of CTDs—the inexperience of the work force—is more controversial. Everyone concedes that the work force is inexperienced. Even those workers who had experience at Big Three auto plants needed time to adjust to the Japanese production process, and workers who had no automotive experience needed even more time. Company officials contend that workers simply need to become "work hardened" and more experienced in their jobs. They point to a decline in CTDs over time where three conditions are met: (1) there is a steady work force; (2) ergonomic job redesign is implemented; and (3) job rotation is used. Nevertheless, as we shall see, conditions one and three have become increasingly difficult to meet because the rate of CTDs is so high.

Critics in the union concede that there are benefits to meeting these conditions but contend that the high rate of CTDs can be attributed to the inadequate training given recent hires and the fast work pace. They argue that managers have moved new workers onto the assembly line before they were completely trained, thereby increasing their risk of injury. Moreover, as efforts were made to find easier jobs for workers with CTDs, there were fewer easy jobs for other workers to rotate through, resulting in greater stress and more disabilities overall. The disparities in the difficulties of work tasks has also contributed to conflict among team members.

As the numbers of injuries and illnesses have mounted, disputes have arisen, leading to grievances and in some cases lawsuits. Some workers with CTDs have reportedly been fired or not called back after an absence if easy jobs were not available. The practice of dismissing disabled workers has been particularly hard on part-timers, who, according to union sources, have a higher rate of injuries and illnesses because of their inferior training and more limited experience.

Disputes have also arisen over claims for workers' compensation. Management has argued that the worker was not incapacitated or that the condition was not work-related. Many of these disputes have ended in lengthy adjudication procedures. Meanwhile, the disabled worker is unemployed and has no compensation.

According to a source who requested anonymity, Jidosha USA started contesting workers' compensation claims at the behest of the insurance company that provides workers' compensation insurance coverage for Jidosha USA.[4] According to this source, insurance company officials recommended accepting any claim for which there was physical evidence of an injury or illness and evidence or witnesses to attest that it occurred at work but to contest claims if these conditions were not met. Most CTD-related claims fail to meet these conditions since they develop incrementally over time.

At the height of its use, job rotation was being practiced by 85 to 90 percent of the work force. But, contrary to expectations, it was found to be negatively related to product quality in that workers did not develop the same degree of expertise they did performing one job. The sacrifice in quality was overlooked when demand was at its peak, but when demand fell off, management became more quality conscious and reduced the percentage of the work force involved in job rotation. About 60 percent still rotate.

With the reduction in the percentage of workers who rotate, complaints have arisen regarding favoritism in assigning jobs. Because of differences in hazardousness and difficulty, job assignments are controversial in any manufacturing facility. Compared with traditional UAW plants, however, Jidosha USA has fewer job classifications and there is less emphasis on seniority. Furthermore, team leaders and unit leaders have more discretion in job assignments. Many workers view their decisions as arbitrary and biased (Parker 1990).

One other source of worker discontent is the heat during the summer. Management has attempted to address this problem by installing ceiling fans but contends the matter is primarily one of comfort rather than safety or health. During one heat wave the Maintenance Department posted a detailed memorandum explaining why the plant was so hot. The principal reason given was that there were several glitches in the plant's sophisticated ventilation system. At least one worker was apparently not convinced by the explanation and had printed the word *bullshit* in large letters across the memorandum. The fact that managers (as well as some union members) work in air-conditioned offices adds to the resentment over this issue and obviously conflicts with the espoused egalitarianism in the plant.

The other complaints have been relatively minor. Workers have complained that the chemicals used in a degreasing operation emitted odors and that there were not enough drinking fountains. Japanese managers and advisers thought there were sufficient numbers of drinking fountains and complained that the American workers drank too much water.[5] The American Jidosha USA engineer who related this story had his own explanation—the Japanese workers do not sweat as much. Such stereotyping suggests that there may be underlying cultural and even racial conflicts.

In published reports (Parker and Slaughter 1988), Japanese personnel at Jidosha USA have won praise from their American colleagues for their willingness to fix malfunctioning machines while they were still powered up. Although such practices demonstrate loyalty and commitment to the company, they are flagrant violations of work safety rules regarding lockout procedures. The safety and health staff found this behavior to be particularly galling. Such practices obviously send the wrong message about the priority of safety relative to production. Such apparent conflicts between Japanese production methods and safety and health concerns has led one UAW safety and health expert to quip that *kaizen* means "take off all the guards, jump into the machine, and the foreman works."[6]

That differences in Japanese and American attitudes with respect to safety and health have the potential to produce conflict became evident one day as the Japanese safety and health adviser acted out the typical reactions of American and Japanese workers to receiving a minor cut. Japanese workers put their cut hand in a glove so no one sees it. American workers complain, "Ooh, ooh! I want to go to the health center. Cut! Cut! Cut!" The adviser seemed to be suggesting that American workers are "sissies" about reporting minor injuries and that Japanese workers are more honorable in hiding them. Safety and health and public relations staff of Jidosha in Japan who reviewed an early version of the manuscript for this book objected to this interpretation and stated that it was "pure conjecture on the author's part." They did not supply an alternative explanation, however.

One situation in which American Jidosha workers are inclined to hide injuries and illnesses can create problems in meeting safety goals. Because of the ostensible importance placed on safety and health and the increasing difficulty finding jobs, trainees or part-timers are likely to hide minor injuries and illnesses for fear of looking bad. These workers are concerned—apparently justifiably so—that having been injured or having contracted an illness will be held against them when their performance during their ninety-day probationary period is evaluated.

Clearly, American Jidosha workers and their union are more assertive than their Japanese counterparts in pursuing safety and health concerns. In a general discussion on cooperation and conflict at the U.S. plant, union officials bridled at the suggestion that management would like the union to be as cooperative and compliant as the Jidosha Workers' Union. They felt there would always be benefits in having a somewhat adversarial relationship with management. The election of union officers from the dissident New Directions movement reflected this belief and set the stage for even more intensely adversarial relations in the future.

The use of work injury and illness rates to determine bonuses for managers produces another conflict, not uncommon in the auto industry and elsewhere, over the treatment to be provided for specific injuries and illnesses. Managers occasionally argue, for example, that certain cuts should be treated with a butterfly bandage rather than a suture. They apparently believe, mistakenly, that the laceration is a "SOSHA-recordable" injury only if it is sutured. There is no evidence that plant medical personnel have modified treatment because of pressure or complaints from management, but treatment is a source of conflict nonetheless. I have heard about similar conflicts in U.S. auto plants, but in these cases it was the safety and health managers who were arguing with the medical personnel (Wokutch 1990a). These conflicts were obviously more serious since they were within the safety and health/medical services system. Nonetheless, the safety and health director at Jidosha USA was concerned that government regulators would view suspiciously the use of injury and illness rates in determining bonuses.

Workers at Jidosha USA believe the union safety and health representative is at least partially responsible for working conditions at the plant. He has been able to maintain good relations with mainstream UAW members and leaders as well as with those from the New Directions movement. He has also been able to maintain generally good rapport with the rest of the safety and health staff, although controversies developed when the plant moved to full-speed production.

When the UAW plant chairman elected on the New Directions slate was initially asked what role safety and health issues had in his election, his immediate response was that they did not play a significant role. On further questioning, however, it became clear that safety and health–related issues such as the work pace, job rotation, production staffing levels, and heat were, in fact, crucial in the election. The initial confusion suggests that safety and health was not a theme in the election but also that labor and management kept safety and health conditions at least somewhat removed from the political controversy.

It is clear that the company would like the union to be more cooperative concerning safety and health matters. The Japanese safety and health adviser complained that the UAW criticizes safety and health conditions in the plant yet, unlike the Jidosha union, does little to improve conditions. He did not cite specific initiatives that he would like the union to undertake, but he would undoubtedly welcome assistance in getting workers to follow work rules more closely. Although the degree of labor-management cooperation on safety and health at Jidosha USA is greater than that at most American-owned auto plants, it is unlikely that it will ever approach the levels in Japan, which suits most workers just fine. Union leaders, however, even those from the New Directions movement, complain about the lack of cooperation by management. They feel that many of the American managers at Jidosha USA are uncomfortable with labor's participation in decision making and attempt to dictate the terms of labor's involvement in workplace issues. Some suggest that the Japanese managers are more supportive of worker participation.

Despite the conflicts and problems, managers, workers, and union officials expressed considerable pride and confidence in the safety and health system at Jidosha USA. The safety and health director conceded that there were problems and that much more needed to be done but noted that he had accomplished more in two years at Jidosha USA than he had in fifteen years with his previous employer. Even after he resigned, he remained an enthusiastic supporter of the Jidosha approach to safety and health management.

U.S. versus Japanese Standards and Practices

As a rule, Jidosha practices conform with SOSHA standards. Several differences were uncovered, however, during the SOSHA officials' tour of Jidosha's facilities in Japan. Both SOSHA and Jidosha officials (Japanese and American) were asked to comment on these differences and to assess the relative merits of conventions in the two countries. Notably, both groups cited far more Japanese standards and practices that they believed were superior. Several reflected a pattern in Japan whereby safety practices and requirements tend to be applied universally rather than on a hazard-specific basis as in the United States. Thus, all workers go through safety and health training, not just those who are expected to perform hazardous work; all workers of a certain age are given medical checkups; and emergency stop buttons are required on all main control and subcontrol panels on machinery.

The standards governing robot guards were also perceived to be superior. SOSHA officials judged them to be more protective than anything even under consideration by SOSHA or federal OSHA. They specifically cited requirements for redundant circuitry to ensure safe operation if circuitry failed. They also mentioned provisions calling for workers to "dress" the tips of the robot through a window rather than by entering a fenced-off area, which could expose them to danger. In addition, they noted that there is a requirement in Japan that a permit be issued whenever hazardous equipment is to be installed (rather than only if construction is involved). One SOSHA official noted that defining what constitutes "hazardous" equipment would be problematic.

SOSHA officials were also impressed by the human aspects of safety and health management. These included the assumption of safety and health responsibilities by line managers and the cooperation on safety and health between labor and management and between government regulators and firms. They thought such practices as having industry personnel train government regulators in the safe use of new technology was particularly impressive. SOSHA representatives judged labor-management relations and government-industry relations to be generally conducive to dealing constructively with safety and health issues.

One example of the differences in labor-management-government relations is telling. Both SOSHA and the Labor Standards Bureau *recommend* the use of joint safety and health committees. For cultural and political reasons, however, the government recommendation carries far more weight and is more likely to be followed in Japan than in the United States.

SOSHA officials cited a few U.S. safety and health practices that they thought were superior to those in Japan. One concerned the conduits used for electrical wiring. In keeping with their emphasis on flexible manufacturing systems, the Japanese prefer flexible conduits, which permit rapid changeovers of equipment. They also contend that rigid conduits are less needed in Japan than in the United States because the electrical current in Japan is weaker. In fact, the current in Japan is strong enough to electrocute someone who comes in contact with a frayed wire. More to the point, the Japanese rely more on workers to follow appropriate procedures and therefore to avoid contact with the wires.

Machine motors in Japan are also wired and secured in such a way as to facilitate more rapid repair or replacement should something malfunction. This is consistent with the flexible Japanese management system and with the philosophy of correcting problems immediately once they are discovered. Still, this is done at the expense of some of the protection afforded by the more secure motor housings required in the United States.

Several differences were noted regarding the use of two-hand control buttons, which are more common in the United States than in Japan. In characteristically understated Japanese criticism, Jidosha officials said that the SOSHA regulations pertaining to these controls were "more severe" than in Japan. In particular, they noted that two-hand controls are required on jig operations even though a worker would not get seriously injured if his or her hands were caught in the machine.

Another difference pertained to the distance between the control buttons. Occasionally, a worker will attempt to circumvent two-hand controls by placing something on one of the controls or jerry-rigging a "bridge" that allows the worker to hit both buttons with one hand. The worker can then use the other hand to do something else, thereby increasing his or her speed. The temptation to sabotage the controls is especially strong if the workers are paid on a piece-rate basis. In the United States, the control buttons are placed farther apart than in Japan to make bridging more difficult. In Japan, workers are trusted not to circumvent mandatory safeguards.

SOSHA officials also had a recommendation for the Japanese regarding the guard rings around stop and start buttons on machines. The officials recommended that the diameter of the rings be enlarged so that workers would hit the control buttons with the palms of their hands rather than just their fingertips. This would reduce the stress on the wrist and the likelihood of workers developing carpal tunnel syndrome, but it would make it easier to sabotage the two-hand control with a bridge. Although carpal tunnel syndrome is not considered as much of a problem in Japan as it is in the United States, Jidosha officials said they would *consider* enlarging the guard rings when they remodeled old facilities or built new ones. They were not sure the Labor Standards Bureau would allow the change, however, even if Jidosha wanted to make it.

The relative lack of concern about carpal tunnel syndrome in Japan appears to result from both perceptual and actual differences. It is only in the last few years that this and other cumulative trauma disorders have been recognized as serious problems in the United States. Even though such disorders existed in the past, although perhaps to a lesser extent, they were not usually diagnosed as such. Japan may be in a similar stage in recognizing the problem. Still, there are reasons to believe that CTD rates are lower in Japan, although not as low as the data suggest.[7] Given the greater homogeneity of the work force in Japan, ideal work motions (i.e., ideal for achieving maximum efficiency and safety and health) can be specified, and, because of the obedience of Japanese workers, they are more likely to be followed.

Another discussion between the SOSHA officials and the Jidosha managers about guarding requirements typified the differences in orientation to worker protection between Japan and the United States. The SOSHA officials were explaining the requirements for a containment system in areas where flammable liquids are used and the Japanese wanted to know why such a system was necessary. The obvious answer, of course, is to prevent the liquid from spreading in the event it is spilled. The predictable response of the Japanese was, "Why would anyone spill it?"

Overall, the SOSHA officials were extremely impressed with the regulation and management of safety and health in Japan and thought that significant benefits could be derived by adopting some Japanese practices. The Americans were particularly impressed by the way concern for safety and health is integrated into the production system. As a result, there is far less need for the government to play the role of police officer to ensure compliance.

The SOSHA officials were so impressed with the status of regulatory relations in Japan that upon their return to the United States they tried to modify some of their own practices to approximate more closely those of the Japanese regulators. For example, senior SOSHA officials instructed their field staff to pay more attention to the effects of regulations on production in an effort to develop more harmonious government-industry relations. A year or so later, however, they said they had not observed any change in relations. Moreover, as the excitement about the visit to Japan declined and problems developed at Jidosha USA, SOSHA officials, like Jidosha workers, became less enthusiastic about safety and health practices at the plant.

In most cases, Jidosha officials thought their practices were superior to SOSHA standards. Nonetheless, they quickly agreed to adopt the U.S. standards in their U.S. operations when they were more stringent. Japanese Jidosha officials remained unconvinced that most of the "more stringent" SOSHA regulations were necessary, and it is unlikely that they had any serious intention of adopting them in their Japanese operations. Jidosha USA follows the standards of Jidosha Japan when they are considered more protective and when they do not violate SOSHA regulations.

Comparative Staffing Levels

Comparing safety and health staffing levels at the Jidosha plants in Japan and in the United States is difficult since there are differences in job descriptions. In Japan, for example, people typically do their own clerical work and no one person has comparable responsibilities to the clerical assistant in the U.S. plant. Further, there is no counterpart in the Safcty and Health Department to the UAW representative. The Jidosha union safety and health specialist is not formally part of the company safety and health unit, although he operates in conjunction with it. In fact, officials at Jidosha in Japan apparently questioned the wisdom of the U.S. structure when it was first proposed and still find it strange. Nor is there a counterpart in Japan to the Japanese safety and health adviser at Jidosha USA.

If one considers all 9 individuals in the Jidosha USA Safety and Health Department full-time members of the safety and health staff, the ratio is 1 staff member for every 333 workers (3,000/9). A more conservative calculation, assuming the UAW representative devotes only one-fifth time to company safety and health issues (based on his estimate that 80 percent of his time is devoted to union safety and health responsibilities), results in a ratio of 1 staff member for every 366 workers (3,000/8.2). These ratios are somewhat lower than the ratio for the entire Jidosha operation in Japan (1 safety and health staff member for every 260 employees)[8] but close to the ratio at the new Jidosha plant in Japan, which has 13 safety and health staff members for approximately 5,000 workers (1:385). This is probably the more meaningful comparison since neither figure includes staff at Jidosha's headquarters.

Compared with safety and health staffing levels in U.S. companies, Jidosha USA's level is very high. As noted in the last chapter, 1 staff member for every 1,750 workers is fairly typical; the ratio at Jidosha USA is almost five times that. The safety and health director pointed out that the staffing level is much higher than it was at his previous place of employment, a modern "showcase" auto plant operated by one of the Big Three. It was approximately the same size as the Jidosha USA plant but had only three safety and health professionals, including himself (giving it a ratio of about 1 staff member per 1,000 workers). These disparities are all the more significant given the greater emphasis on and acceptance of safety and health responsibilities by line managers and workers in Jidosha than in U.S.-owned firms.

Comparative Performance

Although there are a great many difficulties in doing so, it is important to assess the effectiveness of the safety and health system at Jidosha USA relative to that of Jidosha in Japan and of American-owned auto firms. The most telling evidence that members of the Safety and Health Department think the Jidosha system in Japan is superior is that the director of the department repeatedly stated that he was reluctant to submit to an in-depth comparison of the performance and activities of his department relative to those of Jidosha in Japan.

Initially, his stated reason for not wanting to compare Jidosha USA's operations and those of Jidosha in Japan was that the U.S. operation was in the start-up phase of production and the Japanese operations had been in place for years. Large differences remained, however, and in many cases became more pronounced, even after the workers had become more experienced. Thus, other explanations were offered. These centered primarily on cultural differences (Japanese workers are more accustomed to and willing to follow orders regarding appropriate work practices; and Japanese workers, managers, and medical personnel typically do not regard soft-tissue disorders such as CTDs or strains and sprains as legitimate work injuries or illnesses) and physical differences (the Japanese work force is more homogeneous in terms of sex, strength, size, speed, and ability to use both hands, so that engineers could develop production machinery and work practices that closely approximated the ideal for all workers). These explanations certainly made sense and were consistent with my observations.

In spite of their reluctance, members of the Safety and Health Department at Jidosha USA made several comparisons. They said U.S. workers were less careful about following safety and health work rules and sought technological fixes for problems rather than addressing them through their own actions. The Japanese safety and health adviser used as an example a department that handles sheet metal and has a high incidence of lacerations. Despite having been told repeatedly to grip the sheet metal firmly, the workers get cuts from sliding their hands up and down the sharp edges while carrying the metal sheets. In contrast, workers in a similar department in Japan avoid getting cuts by gripping the sheets firmly. Workers at Jidosha USA requested that a machine be installed to eliminate the need to handle the metal. Management was not particularly sympathetic to this request because the solution was obviously more expensive than getting workers to follow directions.

On the whole, in spite of the problems, workers at Jidosha USA are more conscientious about following work rules than workers in the rest of the U.S. auto industry (see Wokutch 1990a). In particular, the emphasis on group activities produces peer pressure to conform. A worker who was not wearing required safety apparel such as a helmet or glasses would probably be gently chided or kidded about it by fellow workers and as a result would probably put on the necessary gear.

Any comparisons of work injury and illness rates between Jidosha operations in the United States and Japan have to be interpreted cautiously because of differences in reporting practices and definitions, as well as the lack of experience of the work force at Jidosha USA. Further, statistics for Jidosha USA are available only for limited periods of time because the facility opened so recently and because Jidosha officials are reluctant to divulge U.S. injury and illness statistics. This reluctance apparently stems from embarrassment about the injury and illness rates and sensitivity about contentiousness in labor-management relations regarding safety and health. The statistics reported here were pieced together from various unofficial but knowledgeable sources at Jidosha USA who requested anonymity. They feared negative repercussions might befall them for having divulged such highly sensitive information. The volatility of the rates reported here coupled with the labor-management controversies over injuries and illnesses at the Jidosha USA plant raises some questions about the validity of the data.

Sources among management contend that some workers, at least early on in production, may have made bogus CTD claims, thereby inflating early injury and illness statistics. Reporting procedures were clearly

Table 5-1. Injury and Illness Experiences of Jidosha USA and Jidosha Japan

Fiscal Year	Jidosha USA	Jidosha Japan	Multiple[b]
Total injury and illness rate (Total injuries and illnesses per 200,000 hours worked)			
1989[a]	32.2	N.A.	N.A.
1988	44.4	.158	281
1987	N.A.	.196	N.A.
Lost-time injury and illness rate (Lost-time injuries and illnesses per 200,000 hours worked)			
1989[a]	5.9	N.A.	N.A.
1988	12.4	.062	200
1987[c]	1.0	.060	17
Severity rate[d] (Lost workdays due to injuries and illnesses per 200,000 hours worked)			
1989[a]	133.0	N.A.	N.A.
1988	80.0	31.8 (6.21)	2.5 (12.9)
1987	N.A.	34.0 (9.27)	N.A.
Cumulative trauma disorder rate (Cases per 200,000 hours worked)			
1989	5.92	N.A.	N.A.
1988	10.10	N.A.	N.A.

Source: Jidosha Japan and Jidosha USA records.
N.A. = not available.
[a]First ten months of fiscal year only.
[b]Jidosha USA injury and illness rate stated as a multiple of the Jidosha Japan rate.
[c]First ten months of operation only.
[d]Figures include days away from work only for the United States; restricted work activity days and charges for fatalities or disabilities are not included. Severity rate figures for Japan also exclude restricted work activity days. The first set of figures for Japan includes charges for both fatalities and disabilities, but figures in parentheses exclude charges for fatalities.
[e]First ten months of fiscal year.

tightened up in fiscal year 1989. The official company position is that the procedures were changed so as to eliminate bogus claims, whereas union officials argue that later on legitimate claims were denied. One management source told me confidentially that this argument was valid.

Keeping the above caveats in mind, reported work injury and illness rates are enormously higher at Jidosha USA than they are in Japan. Table 5-1 shows total incidence rates, lost-time incidence rates, and severity rates for Jidosha USA and Jidosha in Japan. On each of these dimensions the rate for Jidosha USA is stated as a multiple of that for Jidosha in Japan for the respective time period. In addition, information on CTDs is included for Jidosha USA. The total incidence rates at Jidosha USA in fiscal years 1988 and 1989 were 44.4 and 32.2 respectively per 200,000 hours worked. By comparison, the rates were .196 and .158 for fiscal years 1987 and 1988 for all of Jidosha in Japan.[9] Thus, for the one year for which there are common data, 1988, the total incidence rate at Jidosha USA was 281 times higher than at Jidosha in Japan.

Differences in the lost-time incidence rates are on a similar order of magnitude. For the first ten months of operation, when the line was operating well below full speed, the lost-time incidence rate for the Jidosha USA plant was 1.0. The rate increased dramatically to 12.4 when production began at full speed in fiscal year 1988 but declined to 5.9 in 1989. Thus, the lost-time incidence rate at Jidosha USA during the first ten months of operation was seventeen times as high as the rate at Jidosha in Japan for the same fiscal year. In fiscal year 1988, however, the rate at Jidosha USA was an enormous two hundred times higher. And although it was much lower than in 1988, the rate for fiscal year 1989 was still seventy-five times the average rate at Jidosha in Japan for the four previous years (1989 data are not available for Jidosha in Japan).

Severity rates (number of days away from work per 200,000 hours worked) for the U.S. operations are more similar to those of Jidosha in Japan: 80 lost workdays per 200,000 hours worked in 1988 and 133 in 1989, compared with 34 in 1987 and 31.8 in 1988 for Jidosha in Japan. The figures for Jidosha in Japan are inflated, however, because there was a fatality in each of these years. The Japanese recordkeeping convention is to add 7,500 lost workdays for each fatality in calculating the severity rate for the year the fatality occurred. Since no such days are added in calculating U.S. severity rates (there have been no fatalities at Jidosha USA in any event), I have calculated the severity rate for Jidosha in Japan minus the fatality charges.[10] These are 9.27 and 6.21 for 1987 and 1988, respectively, indicating that the severity rates are largely inflated. For 1988, the severity rate for Jidosha USA was 2.5 or 12.9 times that of Jidosha in Japan, depending on whether the fatality charges are included.

In addition to aggregate statistics, I was also given statistics on cumulative trauma disorders. These occurred at Jidosha USA at a rate of 10.1 per 100 full-time workers in fiscal year 1988 and 5.92[11] per 100 full-time workers in fiscal year 1989, when concerted efforts to reduce CTDs were undertaken. Some (perhaps much) of this 41 percent reduction is attributable, however, to the change in injury and illness reporting and recording practices in 1989. At that time, the company began contesting most CTD claims, demanding physical evidence of the existence of a disability before filing a workers' compensation claim or recording the incident as an injury or illness.

CTDs are reported at a far higher rate in Jidosha USA than in Jidosha in Japan. Precise figures for Japan are not available (Jidosha in Japan has not considered CTDs to be a problem and has never bothered tabulating them as a separate injury and illness category), but officials have estimated that the rate is about *one hundred times* higher at Jidosha USA. Reports from other Japanese automotive transplants suggest that this disparity between the transplant and the Japanese parent is typical.[12]

The disparity may be even greater. The Jidosha USA safety and health director speculated that Jidosha in Japan may record no CTDs. He attributed some of the disparity in the data to the greater homogeneity of the work force and to the more strict observance by Japanese workers of designated work practices. He also felt there were significant differences in recordkeeping practices: "I question whether they are willing to count those aches and pains as legitimate injuries. You have to have a broken bone or some blood before it's recorded." He related an incident in which he saw a production worker at Jidosha in Japan wincing with pain every time he bent over at his work station. When he asked his hosts from the Safety and Work Environment Department what was wrong with the worker, their response was "Oh, that's nothing."

Comparing the Jidosha USA injury and illness experience with that of the U.S. automotive industry reveals mixed results. The lost-time injury and illness rate during the first ten months of operation compares very favorably with the experience of the industry as a whole.[13] It is one-sixth the rate reported in 1987 for U.S. auto plants employing more than one thousand employees (5.9 per 200,000 hours worked[14]), in other words, those plants most like Jidosha USA.[15] Following the start of full-speed production, however, the rates compare very unfavorably with those for the rest of the industry (SIC Code 3710). The total incidence rate for 1988 was 44.4 injuries and illnesses per 200,000 hours worked, or 91 percent higher than the rate for the industry (23.3) and 66 percent higher than the rate for auto plants employing one thousand or more workers (26.8).

The relative performance of Jidosha USA during full-speed production was only slightly better with respect to lost-time injury and illness experience. The lost-time injury and illness rate (12.4) was 48 percent higher than the rate for the entire industry (8.4) and 39 percent higher than the rate for establishments employing one thousand or more workers (8.9).

The relative performance of Jidosha USA on the dimension of lost workdays is much better. The rate in 1988 (80) is only 62 percent of the rate reported for U.S. auto establishments with one thousand or more employees (128.8).[16] Thus, although the rates for total injuries and illnesses and lost-time injuries and illnesses were far higher for Jidosha USA than for the rest of the industry, far fewer days were lost. This may be because the injuries and illnesses were less severe or because company policies encourage sick or injured workers to come back to work as soon as possible.

At the time of this writing, industry injury and illness statistics are not available for 1989, but it is possible to speculate on Jidosha USA's relative performance. OSHA has been levying substantial fines for injury and illness underrecording in the industry, and reported rates have consequently been rising sharply in recent years. It is likely that rates will be higher for 1989 (Mirer 1989). Thus, Jidosha USA's total incidence rate of 32.2 for 1989 will probably be only somewhat higher than the industry average, and the lost-time injury and illness rate of 5.9 will almost certainly be substantially lower than the industry average. The rate for lost workdays (133) will probably be lower as well.

The CTD rates at Jidosha USA are also much higher than the reported rates for other large U.S. auto employers. The 1988 rate for motor vehicle and equipment establishments employing one thousand or more workers (SIC Code 3710) is 1.89 per hundred full-time employees. This is almost triple the rate for 1987 (.64)[17], indicating tremendous volatility on this measure. There have been no changes in production processes that could account for an increase of this magnitude. The most obvious explanation is that it is due to changes in reporting practices. Nevertheless, the rate for Jidosha USA for 1988 is more than five times as high as the rate for comparably large auto industry employers for that year. The statistics are not yet available for the entire industry for 1989, but Jidosha USA's 1989 rate is more than three times the 1988 rate for industry establishments with one thousand or more employees. This disparity exists even though the work force at Jidosha USA on average, is younger, and one would expect healthier, than the work force in the rest of the industry.

One must be cautious in making these comparisons, however, in that the accuracy of the CTD rates for the industry is questionable. In addition to the increases cited above, the recorded illness rate (of which CTDs constitute a majority of cases) increased almost tenfold from 1985 to 1987 for automotive establishments with twenty-five hundred or more employees (Mirer 1989). This increase is attributed primarily to recordkeeping changes. In fact, Franklin E. Mirer (1989) estimates that the true rate of CTDs is really one in ten or ten per hundred full-time workers, which would result in a *higher* rate for the industry than the most current rate for Jidosha USA. Jidosha USA's rates are probably affected by these recordkeeping changes as well, especially since the union has protested the recent tightening of procedures for compensating and recording CTDs. Moreover, simple observation of the pace of work at Jidosha USA would lead one to anticipate more of a problem with CTDs here than at other U.S. auto plants. The relative rates of CTDs for Jidosha USA and for the industry bear watching as record-keeping practices stabilize over time.

The disparities in injury and illness rates between Jidosha USA and Jidosha in Japan are even more intriguing. The Jidosha USA safety and health staff, like others who have compared injury and illness rates in major U.S. and Japanese companies, were astonished at the differences. On several occasions, staff members expressed considerable skepticism regarding the accuracy of the statistics for Jidosha in Japan. One of the more colorful comments in this regard was the assertion that the Japanese figures were "bullshit." Nevertheless, the disparities probably contributed to the forced resignation of Jidosha USA's safety and health director.

Several factors were cited in explaining the differences in the rates, but differences in the definitions of a work injury and illness were viewed as a principal factor. A primary difference is that a restricted work activity injury or illness is not considered a recordable incident in Japan. Further, in Japan any work injury or illness for which a physician provides treatment is supposed to be recorded regardless of whether the worker is incapacitated. In contrast, in the United States only injuries or illnesses requiring certain types of treatment (e.g., use of a suture) by a physician must be recorded if the worker is not incapacitated.

The Japanese safety and health adviser noted that there is also inconsistency in the reporting practices in Jidosha in Japan. He stated that the company is "very big" and the safety and health staff cannot ensure that everyone always follows the appropriate definition. Moreover, penalties for misrecording the number of injuries or illnesses are rarely imposed.

Finally, cultural differences contribute to the discrepancies in rates. The difference between the American sense of guilt and the Japanese sense of shame is relevant here. Guilt is internally oriented; it is the feeling of self-reproach one feels for having done wrong. Shame, in contrast, is externally oriented. It is the dishonor or disgrace one experiences when others believe one has done wrong. In Japan it is considered in some sense honorable to hide minor injuries or illnesses since having an accident, regardless of who was at fault, causes workers shame. In contrast, in the United States workers may feel guilty if an accident was their fault but they are not likely to hide their injuries out of shame.

Even taking all of these factors into account, it is difficult to explain the disparities. The obvious conclusion, which is reinforced by observing individual workers and safety and health activities, is that there really are significantly fewer injuries and illnesses in the Jidosha facilities in Japan.

The relatively poor performance of Jidosha USA compared with the rest of the U.S. auto industry is obviously a source of embarrassment and disappointment for its safety and health staff. Their contention is that they are more scrupulous in their recording practices than other U.S. firms and that they record conditions as injuries or illnesses if a worker complains without requiring physical evidence that a problem exists. They claim other firms wait for physical evidence. Although there may be something to this argument, it does not explain the differences entirely. Sprains, strains, and cumulative trauma disorders account for almost 50 percent of all injuries and illnesses at Jidosha USA; CTDs accounted for 23 percent of the total cases of injuries and illnesses in 1988 and 18 percent in 1989. Thus, the figure at Jidosha USA is much higher than at typical American-owned plants. For the U.S. auto industry as a whole, CTDs accounted for only about 12 percent of total injuries and illnesses in 1988 (U.S. Department of Labor, Bureau of Labor Statistics 1990b). The obvious conclusion is that the stresses of the production system do indeed make soft-tissue disorders more of a problem at Jidosha USA. The problem seems to go to the very core of the production system, which elicits maximum efficiency from both workers and machinery. Although accidents must be avoided at all costs in this production system, slow-developing conditions such as CTDs are evidently viewed as less of a threat. Perhaps it is reasoned that workers with conditions such as CTDs can be replaced with only minimal disruption to production. If indeed this is the case, it is a disturbing and short-sighted view and one that is likely to cause increasing problems in the future.

6

Work Injury and Illness Experiences in the United States and Japan

In this chapter I compare work injury and illness experiences in the United States and Japan from several perspectives. I look at historical trends in work injury and illness frequency rates at the national levels for both countries, as well as frequency rates and severity rates in the United States and Japan at the national level, the manufacturing sector level, and the automotive industry level. In addition, work fatality rates are considered at the national level and the manufacturing sector level. I also consider the relationships between establishment size and injury and illness rates for the United States and Japan. Finally, I compare the work injury and illness experiences of automotive firms that have both U.S. and Japanese operations and the experiences of the U.S. operations of Japanese automotive firms with those of U.S.-owned firms. These comparisons reveal the general level of protection accorded workers in the United States and Japan as well as the level accorded specific segments of workers in these countries.

Before considering the statistics, it is instructive to examine the nature of work injury and illness data in the United States and Japan and the social, political, legal, and economic conditions that complicate comparisons. Although they inhibit our ability to make precise comparisons, these conditions tell us a great deal about the national styles of worker protection in these two countries.

Work Injury and Illness Data

Previous research (Wokutch and McLaughlin 1988) suggests that there are nine principal sources of bias in making international comparisons of occupational injury and illness rates: (1) different definitions of work injuries and illnesses; (2) different incentives and disincentives for having an incident officially recorded as a work injury or illness; (3) different methods by which statistics are collected that rely on "reported" or "compensated" injuries and illnesses and that thus exclude certain employers on the basis of size or industrial sector; (4) different bases on which rates are calculated; (5) different national and international political influences regarding the compiling and reporting of statistics; (6) different national traditions regarding reporting; (7) different industrial relations; (8) different distributions by industry of the national work forces; and (9) different degrees of ethnic homogeneity.

Work Injury and Illness Definitions

An important factor in defining work injuries and illnesses is the minimum time a worker must be unable to work before the injury or illness is counted as a lost-work incident. In Japan, the statistics on the *number* of work injuries and illnesses use a criterion of four days. Thus, a worker must be absent for four full days (not including the day of an accident) for an incident to be officially recorded as a lost-work incident. Injury and illness *rate* statistics in Japan refer to injuries and illnesses resulting in an incapacity to work for one or more days (again not including the day of an accident).

In the United States, lost-workday cases are defined in much the same way as they are in Japan; however, lost-workday cases also include restricted work activity cases.[1] The latter are cases in which a worker is able to report to work but unable to perform all the duties of his or her job. This reflects an important difference in the degree of work incapacity necessary for an incident to be considered a lost workday. In Japan, job descriptions are more flexible, so there is no need to record cases involving only restricted work activity as lost-workday incidents. The U.S. Office of Technology Assessment of the U.S. Congress (1985) observed, however, that it is likely that a large percentage of cases involving only restricted work activity go unreported in the United States.

A second major difference is in the practice followed in designating a work incapacity as resulting from either an injury or an illness. In both countries a work incapacity is supposed to be reported as an injury if it is incurred as the result of a one-time event; it is supposed to be reported as an illness if it is caused by repetitive conditions over time. In certain cases, however, such as sore backs, it is often impossible to determine how the condition arose. In the United States, back ailments are supposed to be reported as injuries regardless of whether a single precipitating event can be cited. In other cases, the attending physician or the recordkeeping staff of the safety and health department may have to decide. In Japan, many sore backs or "lumbago" conditions are classified as illnesses in a category called, confusingly, "disease due to hazardous injuries." Because back injuries often account for a large share of the reported injuries in the United States, this difference could substantially affect U.S.-Japanese comparisons of work injury rates *or* work illness rates. The approach I use is to compare work injury and illness rates combined.

A third difference is the inclusion of charges in the Japanese injury and illness severity rate statistics for deaths and permanent disabilities. Such charges are not included in severity rate calculations in the United States.

Reporting Incentives and Disincentives

National standards establishing the duration and extent of work incapacity necessary for an incident to be considered a recordable work injury or illness are set in the political arena. Thus, the relative power of labor and management and their national political influence have a bearing on the determination of these standards. Standards are also likely to be related to various individual and corporate incentives and disincentives to have an incapacity-to-work condition classified as work-related or nonwork-related.

Where there are significant corporate financial incentives to have low reported injury and illness rates, there is greater pressure on managers and, consequently, on workers to report fewer injuries and illnesses. This pressure can take many forms. Workers can be encouraged to ignore minor injuries or illnesses. An injured or sick worker can be assigned legally or illegally to a different and very easy job so that the work incapacity criterion is not met. Plant medical staffs can be encouraged to scrutinize reports of injuries and illnesses to eliminate bogus claims. They can also be pressured to consider the reporting implications of using different medical treatments for an injury or illness. Company injury or illness reports can also be falsified, of course, and legitimate injury or illness claims denied, particularly when workers do not know their rights. At the same time, an individual worker can falsify a claim with respect to its severity or its work-relatedness to get time off from work or better compensation. I found evidence of the existence of all of these practices, although false claims were more often made in Japan so as to *avoid* having a legitimate work injury or illness recorded as such.

Insurance Premiums. There are differences between Japan and the United States in the degree to which firms have a financial incentive to keep reported injury and illness rates low, but the magnitude of the differences is difficult to determine. Workers' compensation premiums in Japan are calculated according to a complex formula that factors in the injury and illness experience of the firm and the industry. In the United States, state laws dictate what insurance is available: self-insurance, state workers' compensation insurance, and/or private workers' compensation insurance. Payouts and premiums vary depending on the insurance, the insurance carrier, and even the size of the insured firm. The closest correlation between payouts and costs to the firm would be expected in firms that are self-insured.

Workers' Benefits. The financial incentives for employees to have an incapacitating condition considered the result of a work injury or illness differ in Japan and the United States. These incentives are related in large measure to the availability and relative generosity of medical coverage and wage-replacement compensation for nonwork-related disabilities. In Japan, workers in virtually all medium- and large-sized companies are covered by medical insurance that provides comprehensive coverage for both work-related and nonwork-related injuries and illnesses. Workers in smaller establishments and others who are not covered by company health insurance are covered by a national health insurance system. The benefits of this system are not as generous, however, as those provided by major company health insurance programs. Thus, it is to the workers' benefit to be eligible for treatment under company health insurance systems.

In the United States, where there is no national health insurance program, the distinction between work- and nonwork-related injuries and illnesses is very important, especially for long-term disabilities. Although both work-related and nonwork-related disabilities are covered under company medical insurance programs (at least in the short run), the coverage is generally more generous for work-related disabilities. An employee who anticipates having long-term effects from an injury or illness is likely to want to have it classified as work-related. From the company's perspective, it is usually more costly if an employee receives workers' compensation rather than company medical insurance benefits.

Researchers in the United States have found a relationship between workers' compensation benefits and work injury reporting (Butler and Worrall 1983). The greater the benefits, the greater the number of reported injuries. In contrast, contests in which work groups with zero injuries (or the fewest injuries) receive prizes have been criticized for creating group incentives and pressures to hide injuries.

In Japan, as in the United States, workers who are disabled are eligible for wage-replacement compensation whether the disability is work-related or not. In both countries, the wage-replacement payments for work-related disabilities come from the employer and the payments for nonwork-related disabilities come from the government in the form of welfare or social security payments. In both cases, the payments from the employer are generally more favorable.

There are many differences within the United States in both medical coverage (usually provided by companies) and wage-replacement compensation (dependent on state workers' compensation laws, company policies, and social security guidelines). There are also differences on these dimensions within Japan.

Both wage continuation and injury and illness reduction are emphasized in the workers' compensation programs in the United States and Japan. Nevertheless, there is considerable question, at least in the United States, whether the employer's financial incentives under the workers' compensation program effectively promote injury and illnesses reduction. Skepticism was a contributing factor, for example, in the passage of the legislation establishing OSHA.

Control Systems. The worker's incentive to have a condition considered work-related often conflicts with the employer's incentives. Where conflicts arise, the nature of the control systems, both corporate and government, for verifying that a condition is work-related can have an important influence on the compensated and reported injury and illness rates. OSHA penalties on U.S. firms for underreporting injuries and illnesses have undoubtedly discouraged this practice. In contrast, in Japan, penalties are rarely imposed for underreporting. Moreover, Japanese workers appear to have their own internal controls that inhibit them from reporting many injuries and illnesses. As a result, Japanese work injury and illness data may substantially understate actual experiences.

Methods of Data Collection

There are two basic national recordkeeping approaches to counting work injuries and illnesses. Reported cases are based on national compulsory injury and illness reporting systems. Compensated cases are derived from government or industry compensation data.

Reported Cases. In the United States, national work injury and illness statistics are derived from annual surveys of establishments conducted by the U.S. Bureau of Labor Statistics. Establishments are selected randomly for inclusion in the survey sample so as to reflect overall employment in the economy. In recent years, the federal government has collected additional injury and illness data, using several supplementary approaches.

In Japan, statistics on injury, illness, and fatality *rates* are collected through surveys conducted by the Ministry of Labor. Statistics on the *number* of work fatalities come from a separate reporting system. When a work fatality occurs, it is reported to the local Labor Standards Bureau or Office, which then reports it to the Tokyo headquarters of the Ministry of Labor.

Compensated Cases. In the United States, injuries and illnesses compensated through state workers' compensation programs are not tabulated on a national basis. Such a tabulation would be extremely difficult given the differences among the fifty state workers' compensation programs.

In addition to survey data, Japan collects statistics on injuries and illnesses compensated through its national workers' compensation system. These statistics refer to the *number* of "casualties" resulting in four or more days' absence from work. In Japan, the incentives for employers and employees (or survivors of employees) to have, or not to have, an injury or illness classified as work-related directly affects the statistics. By contrast, the effect in the United States is indirect. The standards for reporting work injuries and illnesses to OSHA and to state workers' compensation programs are different; however, the determination that an incapacity-to-work condition is work-related for one system will usually influence its designation under the other system.

Size and Industrial Sector. Another difficulty in making international comparisons arises from differences in coverage with respect to employer size and the industrial sectors included. In the United States, nearly all private sector establishments with one or more employees are included in published injury and illness statistics. The exception is agricultural production establishments (i.e., farms), of which only those with eleven or more employees are included. Published occupational fatality data generally exclude all establishments with ten or fewer employees.

The Japanese statistics on the number of work fatalities and casualties resulting in four or more days' absence from work cover private sector establishments with one or more employees. The statistics on fatality, injury, and illness rates cover virtually all private sector establishments with one hundred or more employees. Injuries and illnesses to seamen are excluded from statistics on numbers of injuries and illnesses in Japan. Workers in the construction industry, a notoriously hazardous industry both in Japan and the United States, are excluded from Japanese injury and illness rate statistics.

Bases for Injury and Illness Rate Calculations

The varying bases on which injury and illness rates are computed constitute another problem in making international comparisons. Worker fatality rates in the United States are generally reported on the basis of 200 million hours worked, whereas in Japan they are reported based on 1 million hours worked. The nonfatal U.S. injury and illness statistics reported here are given in terms of injuries and illnesses per hundred full-time workers, while the Japanese statistics, like fatality rates, are reported in terms of occurrences per million hours worked. The standard assumption that a full-time worker in the United States works 2,000 hours (and that one hundred full-time workers work 200,000 hours) allows us to convert from one basis to another.

Political Considerations

The political significance of injury and illness statistics within countries and within the international community of countries further complicates international comparisons. For example, different political factions in the United States attempt to use increases or decreases in reported injury and illness rates to their political advantage and to support their preconceived ideas. Efforts are also undertaken to develop statistics that support a group's ideas. OSHA's counting of "records inspections" is one example (Wokutch 1990a).

Consistent with what I called the myth of the abused Japanese worker, Japan is occasionally accused by trading partners of overworking and abusing workers to gain a competitive advantage. Japanese government, industry, and, in some cases, union officials are concerned that unfavorable reports on occupational safety and health conditions may result in further pressures on Japan to restrict exports or encourage imports. Such pressure could affect injury and illness reporting practices.

Industrial Relations

The nature of labor-management relations can also have an impact on reported injury and illness rates. When relations are adversarial, workers may be more inclined to report minor injuries and illnesses to gain negotiating leverage, to encourage support for union organizing campaigns, to "get back at" management (or a particular manager) for some disliked action or policy, or to protest the pace of work. In such an atmosphere, managers may pressure the medical staff to scrutinize carefully workers' reports of injuries and illnesses to screen out "phony" claims. They may view workers as lazy and reports of minor injuries or nonverifiable illnesses (e.g., a "sore back") as attempts to avoid work; they may be unconcerned with the safety and health of the work force beyond their effects on production; or they may try to establish a negotiating position. Adversarial labor-management relations and their attendant effects on reporting are much more common in the United States than in Japan.

Japanese firms operating in the United States fall in between U.S. firms in the United States and Japanese firms in Japan on this dimension, although relations at Jidosha USA are much closer to those at traditional U.S. firms. Where labor-management relations are more cordial, workers may not exaggerate injuries and illnesses but, rather, may hide them out of a sense of (misguided) loyalty to their superiors.

National Traditions

Differences in national and corporate traditions can make comparisons difficult as well. Organizational commitment, group loyalty, or even fear may deter workers from reporting injuries and illnesses. In Japan, zero-accident campaigns,[2] coupled with cultural traditions of loyalty to one's work group and one's organization and the Eastern sense of honor or "face," lead to an underreporting of injuries and illnesses. Having one work injury or illness makes the zero-accident goal unattainable and is regarded as failing in one's responsibilities to the group. Thus, there is strong internalized pressure not to report injuries or illnesses.

In contrast, in the United States, attempts to "beat the system" are considered acceptable behavior in many circles, and workers fake injuries or illnesses to get a "paid vacation." Different attitudes toward obeying the law and toward working while in pain also influence the statistics.

Industrial Composition

One would expect countries with relatively more workers in high-risk industrial fields such as mining, construction, and manufacturing to have higher injury and illness rates than countries in which low-risk service industries dominate. Moreover, there are significant differences in risk even within industrial sectors. For example, the U.S. Department of Labor, Bureau of Labor Statistics (1990b) reports that the total injury and illness incidence rate for all U.S. manufacturing firms in 1988 was 13.1 cases per hundred full-time workers. The range, however, was from 1.9 cases per hundred full-time workers in printing trade services to 36.3 cases per hundred full-time workers in ship and boat building and repairing. Such differences are almost certainly due to differences in the inherent danger of the work.

Relative to the United States, a higher percentage of Japan's work force is employed in the higher-risk industrial sectors of manufacturing and agriculture, forestry, and fishery. In the United States, a higher percentage is employed in the lower-risk service and government sectors. A higher percentage is also in the relatively hazardous mining industry, however. In addition, the exclusion of the construction industry from Japanese national injury and illness statistics affects comparisons in that the incidence of injuries in that industry is probably high.

Ethnic Homogeneity

The homogeneity or diversity of the national work force is also likely to distort comparisons. Where there are large numbers of foreign workers, one might expect higher injury and illness rates. In many cases, foreign workers do not speak the national language, are less educated, are less trained in their jobs, and are unaware of their rights in the workplace. Further, foreign workers may be reluctant to refuse to do dangerous work for fear of job loss and deportation or other negative consequences. Gloria L. Lee and John Wrench (1980) found that foreign workers in Great Britain had higher injury rates because they were disproportionately represented in more hazardous jobs. At the same time, the factors that contribute to their having more injuries and illnesses also lead foreign workers to hide their injuries and illnesses to a greater extent. By contrast, legal immigrants to Japan may be more inclined than native Japanese to report legitimate injuries and illnesses.

The ethnic compositions of Japan and the United States differ enormously. The United States has a very high percentage of foreign-born workers, including naturalized citizens, resident aliens authorized to work, and undocumented aliens. This last group is especially problematic with respect to occupational injury and illness statistics. Although there are certainly large numbers of such workers, precise statistics on them are, by definition, difficult to obtain. They are probably overrepresented in low-paying, hazardous jobs, such as agricultural field work, that U.S. citizens shun. It is also reasonable to assume that they do not report many of their injuries and illnesses because they are afraid of being deported or because of pressure from employers.

Japan has the most homogeneous population of any industrialized nation in the world because of the tight control it has had on immigration throughout its history. Relatively few foreigners have been allowed into Japan to work, and the population of illegal immigrant workers has been tiny. Both legal and illegal immigration are increasing rapidly (albeit from a tiny base) as the Japanese economy becomes more international and as foreign workers are attracted to the difficult manual labor jobs that many Japanese now avoid. Nonetheless, the numbers of work injuries and illnesses to immigrants is insignificant with respect to national totals.

National Work Injury and Illness Experiences

Trends

Table 6-1 provides longitudinal statistics on the numbers and rates of occupational injuries and illnesses for the United States and Japan. Because of the aforementioned differences, it is not possible to make meaningful comparisons between these two countries, although the statistics shed some light on long-term trends.

The U.S. figures provide good examples of both the complexity of the injury and illness reporting system and how differences in reporting practices can occur over time. Before 1970, occupational injury data were of limited usefulness. Before then, the U.S. Bureau of Labor Statistics relied on voluntary reporting of injuries, and not all industries were covered. The work injury reporting standards were also different than those used today (U.S. Congress, Office of Technology Assessment 1985). Moreover, as the Office of Technology Assessment notes, assessing the extent of occupational illnesses has historically been even more difficult, and these data continue to be less accurate today. The current system of data collection was mandated by the 1970 Occupational Safety and Health Act.

The U.S. injury and illness statistics reported in table 6-1 show an increase from 3.3 lost workday injury and illness cases per hundred full-time workers in 1972 to 4.3 in 1979, followed by a decline to 3.4 in 1983 and then, with some yearly fluctuations, an increase to 4.0 in 1988. The overall increase from 1973 to 1988 was 21 percent. This apparent increase may be a spurious recordkeeping artifact, however. Some supporters of OSHA suggest that significant underreporting of injuries and illnesses occurred during the years after the establishment of OSHA because of confusion about recording requirements. As people learned more about these requirements, the number of *recorded* injuries and illnesses increased. In addition, the increases in 1987 and 1988 are very likely because of an increase in employer conscientiousness in response to the large fines imposed by OSHA for underrecording. Long-term National Safety Council statistics demonstrate a substantial U.S. decline in injury and illness rates over the century. These data may represent a more accurate picture of U.S. trends than the shorter-term Bureau of Labor Statistics (BLS) data, which are greatly affected by changes in recording practices.

The Japanese statistics on the number of injuries and illnesses come from workers' compensation data and cover private sector establishments with one or more employees.[3] The rate statistics include all private

Table 6-1. Work Injury and Illness Experiences in the United States and Japan

	United States		Japan	
	Number of Injuries and Illnesses[a] (in thousands)	Injuries and Illnesses per 100 Full-Time Workers[a]	Number of Injuries and Illnesses[b] (in thousands)	Injuries and Illnesses per Million Man Hours[c]
1958	N.A.	N.A.	401.760	N.A.
1959	N.A.	N.A.	435.017	N.A.
1960	N.A.	N.A.	468.139	N.A.
1961	N.A.	N.A.	481.686	N.A.
1962	N.A.	N.A.	466.126	N.A.
1963	N.A.	N.A.	440.547	N.A.
1964	N.A.	N.A.	428.558	N.A.
1965	N.A.	N.A.	408.331	N.A.
1966	N.A.	N.A.	405.361	N.A.
1967	N.A.	N.A.	394.627	N.A.
1968	N.A.	N.A.	386.443	11.08
1969	N.A.	N.A.	382.642	10.37
1970	N.A.	N.A.	364.444	9.20
1971	N.A.	N.A.	337.421	8.14
1972	N.A.	3.3	324.435	7.25
1973	1,908.0	3.4	387.742	6.67
1974	2,001.8	3.5	347.407	5.11
1975	1,827.5	3.3	322.322	4.77
1976	1,978.8	3.5	333.311	4.37
1977	2,203.6	3.8	345.293	4.32
1978	2,492.0	4.1	348.826	3.91
1979	2,757.7	4.3	340.731	3.65
1980	2,539.9	4.0	335.706	3.59
1981	2,457.5	3.8	312.844	3.23
1982	2,182.4	3.5	294.319	2.98
1983	2,182.7	3.4	278.623	3.03
1984	2,501.5	3.7	271.884	2.77
1985	2,537.0	3.6	257.240	2.52
1986	2,590.3	3.6	246.891	2.37
1987	2,801.6	3.8	232.953	2.22
1988	2,975.9	4.0	226.318	2.09

Sources: U.S. data: U.S. Department of Labor, Bureau of Labor Statistics 1977, 1982, 1987, 1988, 1989, 1990; U.S. President, 1984, 1987, 1988. Data originate from surveys conducted by the U.S. Bureau of Labor Statistics. Japanese data: Japan Institute of Labour 1982, 1988; Japan Industrial Safety and Health Association 1988, 1989. Original sources of data are workers' compensation records for numbers of injuries and illnesses and the Ministry of Labor survey for rates of injuries and illnesses.

N.A. = not available.

[a]Reported injuries and illnesses—lost-workday cases. Coverage—private sector employers with one or more employees.

[b]Compensated injuries and illnesses—through 1972, deaths, injuries, and illnesses resulting in the inability to work eight or more days; from 1973 onward, four or more days. Coverage—all establishments with one or more employees.

[c]Reported injuries and illnesses—injuries and illnesses resulting in an incapacity to work of one or more days (not including the day of the accident for injuries). Coverage—all establishments with one hundred or more employees.

Figure 6-1. Frequency Rate in Japan and the United States (All Industries)

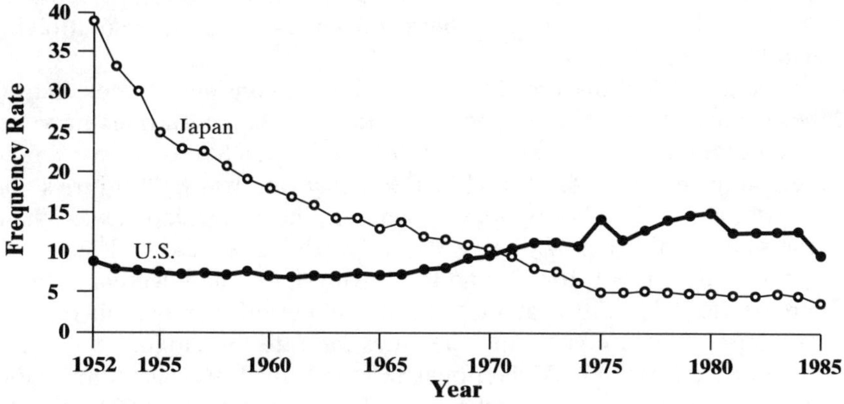

Source: Japan Institute of Labour 1988:13. Ministry of Labor survey; National Safety Council survey.

sector establishments with one hundred or more employees and are based on surveys conducted quarterly by the Labor Standards Bureau. Except for the workers noted above, virtually all members of the private sector are included.

The Japanese system for reporting the number of injuries and illnesses underwent a change in 1973, complicating longitudinal comparisons. Before then, data were available only for injuries and illnesses resulting in an inability to work for eight or more days. Since then, the data have covered injuries and illnesses resulting in an inability to work for four or more days.

There has been a very impressive long-term decline in the reported rate of injuries and illnesses in Japan. In 1968, there were 11.08 injuries and illnesses per million hours worked. By 1972, the year the Industrial Safety and Health Law was enacted, the rate had decreased to 7.25, a drop of 35 percent. By 1988, the rate had fallen to 2.09, representing a decline of 81 percent over the 1968–88 period. Thus, for the 1972–88 period, U.S. Bureau of Labor Statistics data showed a 22 percent increase in U.S. rates, whereas the Japanese Labor Standards Bureau data showed a 71 percent decrease in Japanese rates.

National Safety Council data for the United States provide a better idea of the magnitude of the decline in Japan. The Japan Institute of Labour (1988) has graphed both the industrial injury and illness frequency rates reported by the National Safety Council (fig. 6-1) and the frequency rates for Japan. (National Safety Council incidence rate data are normally based on injuries and illnesses per 200,000 hours worked. They have been converted to million hours worked simply by multiplying by a factor of five.)

Several matters discussed below need to be considered in evaluating these statistics, but there is no doubt that the Japanese rates have declined dramatically relative to those of the United States. According to these statistics, in 1952, the U.S. frequency rate was 8.40 injuries and illnesses per million hours worked, whereas the rate in Japan was 39.24, almost five times the U.S. rate. By 1985, the U.S. rate had increased by about 18 percent, to 9.9, and the Japanese rate had decreased by 94 percent, to 2.52. Thus, at the end of this period, the overall rate for the United States was about four times the rate for Japan.

Despite efforts to make National Safety Council statistics comparable to Japanese statistics, several factors distort comparisons. For one, in 1977, the National Safety Council changed its method for calculating injury and illness rates and adopted OSHA injury and illness definitions. Because of this change, the council recommends against comparing statistics before and after that year as has been done in figure 6-1. Further, council data are reported voluntarily by firms that belong to the organization. It is thus a self-selected and probably biased sample. In contrast, the Japanese data come from surveys conducted by the Ministry of Labor, which selects a random sample of companies that employ one hundred or more workers. Responses are not voluntary. Had BLS frequency data been used for the comparison, however, the U.S. data would have been even less favorable. Using the same conversion procedure, the 1985 rate (3.6 lost workday injuries and illnesses per one hundred full-time workers) (U.S. Department of Labor, Bureau of Labor Statistics 1987) would convert to 18 cases per million hours worked, which is more than seven times higher than the Japanese rate for the same year (2.52). BLS statistics cover establishments of all sizes, however, including those with fewer than one hundred employees. Smaller establishments, which are likely to have a higher incidence of injuries and illnesses, are excluded from the Japanese statistics, resulting in an artificially low reported rate.

Several other factors should also be taken into account. First, field research suggests that there is significant underrecording of work injuries and illnesses in Japan. Furthermore, it is likely that underreporting has increased in recent years as companies and workers have tried to improve on previous years' performances and as greater emphasis has been put on national zero-accident campaigns. Second, unlike the U.S. statistics, the Japanese statistics do not include days of restricted work activity. In 1987, injury cases involving restricted work activity accounted for 11 percent of reported injuries in the U.S. private sector (U.S. Department of Labor, Bureau of Labor Statistics 1989). Third, and finally, because job descriptions in Japan are flexible, it is easy to transfer an injured or ill worker to a less demanding job and not count the injury or illness as a lost-work case.

Work Injury and Illness and Fatality Rate Comparisons

Tables 6-2, 6-3, and 6-4 contain adjusted work injury and illness rate and work fatality rate data for the United States and Japan for the entire private sectors, for the manufacturing sectors, and for the automotive industries. U.S. statistics have been adjusted in several ways using unpublished BLS data to make them more closely conform with statistics for Japan. First, because Japanese statistics cover only establishments with one hundred or more employees, whereas published U.S. statistics cover establishments with eleven or more employees, the statistics for U.S. establishments with eleven to ninety-nine employees have been eliminated. Second, because Japanese injury and illness statistics do not include restricted work activity cases, lost-workday cases involving restricted work activity have also been omitted.[4] Third, the severity rates have been adjusted to follow Japanese practices more closely. Thus, charges of seventy-five hundred days for each fatality have been added to the U.S. figures for days lost from work because of injuries and illnesses. Unfortunately, data were not available on permanent disabilities, so the U.S. severity rates could not be adjusted to reflect charges for permanent disabilities as is done in Japan.

For the sake of comparison, Japanese injury and illness incidence (frequency) rates, severity rates, and fatality rates have been converted to conform with U.S. bases for rate reporting. Thus, published Japanese incidence rates that are reported on a per million hours worked basis have been divided by a factor of five to convert them to a 200,000 hours basis. Likewise, published Japanese severity rate data that are based on a thousand hours worked are converted to the one hundred full-time worker basis used in the United States. Finally, published Japanese fatality rate data that are based on 1 million hours worked are converted to the 200 million hours used in the United States.

The data reveal some fascinating differences. In general, the injury and illness incidence rates in the United States are dramatically higher for each of the three sectors. The severity rates are considerably higher for each of the three sectors, although the disparities are not as great as they are for the incidence rates, and the fatality rates show mixed results. The U.S. automotive industry is the only industrial sector that

Table 6-2. Adjusted Work Injury and Illness Experiences of the Private Sector in the United States and Japan

	United States[a]			Japan[e]		
	Injury and Illness Incidence Rate[b]	Severity Rate[c]	Fatality Rate[d]	Injury and Illness Incidence Rate[f]	Severity Rate[g]	Fatality Rate[h]
1987	3.9	99.61	3.8	.44	40	2.0
1986	3.8	93.47	3.7	.47	44	2.0
1985	3.7	97.90	4.4	.50	58	4.0
1984	3.8	96.15	4.3	.55	68	6.0
1983	3.6	90.18	4.1	.61	60	4.0
Average (1983-87)	3.76	95.46	4.06	.51	54	3.6

Sources: U.S. data: U.S. Bureau of Labor Statistics unpublished statistics. Japanese data: Japan Industrial Safety and Health Association 1984, 1985, 1986, 1988a, 1988b, and 1989 as well as the *1985/1986 Annual Report* and the *1988/1989 Annual Report*. The 1986 fatality rate statistic comes from the International Labour Office 1988.

N.A. = not available.

[a]Reported injuries—lost-workday cases; cases involving only restricted work activity are excluded. Coverage—private sector establishments with one hundred or more employees.

[b]Number of lost-time injuries and illnesses per hundred full-time workers.

[c]Days away from work due to injuries and illnesses per hundred full-time employees. Charges of 7,500 lost workdays due to each fatality are added to U.S. data to enhance comparability with Japanese data.

[d]Fatalities per 200 million hours worked (or 100,000 full-time workers).

[e]Reported injuries—lost-workday cases. Coverage—private sector establishments (excluding construction) with one hundred or more employees.

[f]Number of lost-time injuries and illnesses per 200,000 hours worked, which is equivalent to the working time of one hundred full-time workers.

[g]Days away from work per 200,000 hours, which is equivalent to the working time of one hundred full-time workers.

[h]Fatalities per 200 million hours worked.

has a lower reported fatality rate than its Japanese counterpart, but, because they are derived from relatively small numbers of establishments, the data for this comparison are also the least reliable. In the United States, fatality rates for specific industries are not normally published because of the small sample sizes.

In the U.S. private sector, the 1983–87 average injury and illness incidence rate (3.76 injuries and illnesses per hundred full-time workers) is more than seven times as high as the average rate for the Japanese private sector (.51) for the same five years (table 6-2). In the manufacturing sector, the U.S. rate in those five years (3.7) is approximately eleven times that of the Japanese rate (.34) in the same period (table 6-3). For the automotive industry, the ratio of the U.S. incidence rate to that of Japan is approximately fourteen to one (3.1 versus .22 injuries per hundred full-time workers) for these five years (table 6-4).

Efforts have been made to enhance the comparability of the severity rate statistics by adding charges for fatalities, but the exclusion of charges

Table 6-3. Adjusted Work Injury and Illness Experiences of the Manufacturing Sector in the United States and Japan

	United States[a]			Japan[e]		
	Injury and Illness Incidence Rate[b]	Severity Rate[c]	Fatality Rate[d]	Injury and Illness Incidence Rate[f]	Severity Rate[g]	Fatality Rate[h]
1987	4.0	101.55	3.5	.30	30.0	2.0
1986	3.7	96.56	2.9	.32	36.0	2.0
1985	3.6	91.65	3.4	.33	38.0	2.0
1984	3.7	87.63	3.1	.36	40.0	2.0
1983	3.5	87.03	3.5	.39	44.0	2.0
Average (1983-87)	3.7	92.88	3.28	.34	37.6	2.0

Sources: U.S. data: U.S. Bureau of Labor Statistics unpublished statistics. Japanese data: Japan Industrial Safety and Health Association 1984, 1985, 1986, 1988a, 1988b, and 1989 as well as the *1985/1986 Annual Report* and the *1988/1989 Annual Report.* The 1986 fatality rate statistic comes from the International Labour Office 1988.

N.A. = not available.

[a]Reported injuries—lost-workday cases; cases involving only restricted work activity are excluded. Coverage—manufacturing establishments with one hundred or more employees.

[b]Number of lost-time injuries and illnesses per hundred full-time workers.

[c]Days away from work due to injuries and illnesses per hundred full-time employees. Charges of 7,500 lost workdays due to each fatality are added to U.S. data to enhance comparability with Japanese data.

[d]Fatalities per 200 million hours worked (or 100,000 full-time workers).

[e]Reported injuries—lost-workday cases. Coverage—manufacturing establishments with one hundred or more employees.

[f]Number of lost-time injuries and illnesses per 200,000 hours worked, which is equivalent to the working time of one hundred full-time workers.

[g]Days away from work per 200,000 hours, which is equivalent to the working time of one hundred full-time workers.

[h]Fatalities per 200 million hours worked.

for permanent disabilities in the U.S. statistics creates a downward bias. Nonetheless, the U.S. severity rates are higher than those for Japan for all three sectors considered for all the years for which data are available. For the entire private sector for 1983–87 the average U.S. severity rate (95.46) was 77 percent higher than the average for Japan (54) (table 6-2). For the manufacturing sector, the U.S. average for these five years (92.88) was 147 percent higher than the average rate (37.6) in Japan (table 6-3). The 1983–87 average U.S. severity rate in the automotive industry (65.69) was 113 percent higher than the average rate (30.8) for Japan for the same period (table 6-4).

Although the U.S. severity rates are considerably higher than those for Japan for comparable sectors, the differences are not nearly of the same order of magnitude as the differences in incidence rates. This seems suspicious but may be because I have not been able to adjust the U.S. data to account for the charges for permanent disabilities. It is conceivable (but not likely) that if such charges were added, there would

Table 6-4. Adjusted Work Injury and Illness Experiences of the Automotive Industry in the United States and Japan

	United States[a]			Japan[e]		
	Injury and Illness Incidence Rate[b]	Severity Rate[c]	Fatality Rate[d]	Injury and Illness Incidence Rate[f]	Severity Rate[g]	Fatality Rate[h]
1987	4.2	85.9[i]	N.A.	.19	24	2.0
1986	3.2	79.43	2.4	.20	22	0.0
1985	2.7	54.38	.5	.20	34	2.0
1984	3.1	54.94	.6	.23	26	2.0
1983	2.3	53.8	1.5	.29	48	2.0
Average (1983-87)	3.1	65.69	1.25	.22	30.8	1.6

Sources: U.S. data: U.S. Bureau of Labor Statistics unpublished statistics. Japanese data: Japan Industrial Safety and Health Association 1984, 1985, 1986, 1988a, 1988b, and 1989 as well as the *1985/1986 Annual Report* and the *1988/1989 Annual Report.*

N.A. = not available.

[a]Reported injuries—lost-workday cases; cases involving only restricted work activity are excluded. Coverage—automobile industry establishments (SIC Code 3710) with one hundred or more employees.

[b]Number of lost-time injuries and illnesses per hundred full-time workers.

[c]Days away from work due to injuries and illnesses per hundred full-time employees. Charges of 7,500 lost workdays due to each fatality are added to U.S. data to enhance comparability with Japanese data.

[d]Fatalities per 200 million hours worked (or 100,000 full-time workers); injury fatalities only. Fatality data at the three-digit industry level are not considered reliable and are not normally reported.

[e]Reported injuries—lost-workday cases. Coverage—automobile and accessories manufacturing establishments with one hundred or more employees. The Japanese industrial breakdown "automobile and accessories manufacturing" is composed of automobile manufacturing (including tricycles and bicycles), automobile bodies manufacturing, and automobile parts and accessories manufacturing.

[f]Number of lost-time injuries and illnesses per 200,000 hours worked, which is equivalent to the working time of one hundred full-time workers.

[g]Days away from work per 200,000 hours, which is equivalent to the working time of one hundred full-time workers.

[h]Fatalities per 200 million hours worked (or 100,000 full-time workers).

[i]The 1987 severity rate does not include charges for fatalities.

be the same general magnitude of difference as there is for the incidence rates. A comparison of fatality rates provides even more convincing evidence that there is something curious underlying the relative performance of Japan on these measures.

For the years 1983–87, the average work fatality rate for the private sector was only slightly higher for the United States than for Japan. The average U.S. rate (4.06 fatalities per 200 million hours worked) was only 13 percent higher than the rate in Japan (3.6).

By contrast, the U.S. rates for the manufacturing sector are substantially higher. The average rate (3.28 fatalities per 200 million hours worked) for the 1983–87 period was 64 percent higher than the rate for Japan (2.0) for the same period (table 6-3).

The fatality rates for the automotive industries reveal a different pattern. They are higher for Japan (table 6-4). These figures are of dubious accuracy, however, because of the small sample size.

Two trends are discernible. First, U.S. work fatality rates are comparatively much lower than its incidence rates. Second, the United States generally fares worse for the manufacturing sector than it does for the entire private sector. This is logical. Accidents are extreme examples of inefficiency that greatly conflict with the ideals of the Japanese manufacturing management system. There is thus a strong emphasis in manufacturing enterprises in Japan on reducing accidents and resulting injuries. Although I have not studied other sectors of the Japanese economy, it is possible that worse performance (or just less outstanding performance) in other industries would offset some of the advantage Japan has when rates for only the manufacturing sectors are compared.

The first trend does not have a logical explanation if the data are accurate. Because work fatalities are likely to be investigated closely, they are more likely to be reported accurately than nonfatal injuries and illnesses. The difference between injury and illness rates and fatality rates therefore suggests that nonfatal injuries are underreported to a greater extent in Japan than in the United States.

Undoubtedly, there is underrecording in both countries. After OSHA imposed massive fines on auto firms for underrecording, according to the BLS data, there was a 31 percent increase from 1986 to 1987 in recorded lost-workday injury and illness rates in the automotive industry and an additional 27 percent increase from 1987 to 1988. That these increases were due mainly to changes in recording practices is generally accepted. This development makes the relative degree of underrecording in the United States and Japan that much more intriguing.

One way to attempt to assess the relative degree of underrecording in the two countries is by considering the ratios of injuries and illnesses to fatalities. There are reasons to believe these ratios are meaningful and in some sense predictable. In the early 1940s, Herbert W. Heinrich (1941) observed a stable relationship of 300 to 29 to 1 among the numbers of accidents resulting in no injuries, in minor injuries, and in major injuries.[5] Similarly, military strategists project a relatively stable relationship between wounded and fatally wounded soldiers in combat. Although the ratio of nonfatal to fatal injuries would probably be different in the workplace, one would expect to see a reasonable degree of stability in the percentages, even allowing for some decline in fatalities because medical care has improved. Fujimoto (1967) found, however, that in Japan the rate of fatal injuries to the rate of lost-work injuries *increased* substantially over time. This has led him to suggest (1986) that nonfatal injuries are underrecorded in Japan. Personal calculations show that the ratio of the number of fatalities to the number of injuries and illnesses has also increased in recent years. For example, in 1979, of every 111 casualties reported that resulted in four or more days of lost work, 1 was a fatality; in 1989, the figure was 1 in 90 (Japan Industrial Safety and Health Association 1990).

Because the definitions used in recording injuries and illnesses are so similar in the United States and Japan, one would expect the ratios of nonfatal injuries and illnesses to fatalities to be comparable, assuming roughly equal underrecording of nonfatal and fatal cases. Restricted work activity cases are recordable in the United States but not in Japan; however, there is considerable underrecording of such cases in the United States (U.S. Congress, Office of Technology Assessment 1985) and they should not affect the ratios greatly.

Using the private sector injury and illness and fatality data in table 6-2 leads to the following rough calculations.[6] The average work injury and illness rate for the U.S. private sector for 1983–87 was 3.76 per hundred full-time workers (200,000 hours worked), or 3,760 cases per 200 million hours worked. The average fatality rate for the same years was 4.06 per 200 million hours worked. Thus, the ratio of lost-work cases to fatalities is 3,760/4.06 or 926 to 1. The ratio for Japan during the same period was 510/3.6 or 142 to 1. Thus, the ratio for the lost-work incidence rate to the work fatality rate is approximately 6.5 times higher in the United States than in Japan (926 ÷ 142 = 6.52). Looked at inversely, the proportion of fatalities to total lost-work cases is about 6.5 times as great in Japan as in the United States. Moreover, this discrepancy is increasing because of the continuing increase in *recorded* nonfatal work injuries and illnesses for the United States.

It is possible that factors other than underrecording in Japan account for the disparity. One possibility is that there is greater accuracy in recording fatalities in Japan than in the United States. Evidence has surfaced, for example, that some work fatalities in the United States go unrecorded (Burrough and Lubove 1986). In addition, the U.S. Department of Labor, Bureau of Labor Statistics (1990b) specifically cautions that fatalities are difficult to measure in establishment surveys and that they may be significantly understated in the United States. I am not aware of research in Japan on the extent, if any, of underrecording of work fatalities. If underreporting of fatalities in the United States were the only explanation for these discrepancies, however, it would mean that the level here was actually 6.5 times higher than the reported level, which seems unlikely. Underreporting in Japan to a much greater extent than in the United States appears to be the most reasonable explanation for the disparities.

In spite of the apparent underreporting in Japan, work fatality rates in

Table 6-5. Occupational Injury and Illness Rates by Establishment Size (Number of Workers) for the Manufacturing Sectors in the United States and Japan

	United States[a]				Japan[b]				
	100 or more	100–499	500–999	1,000 or more	100 or more	100–299	300–499	500–999	1,000 or more
1987	3.8	5.4	3.7	2.3	.298	.558	.282	.192	.074
1985	3.6	5.2	3.5	1.8	.334	.624	.350	.222	.098
1984	3.7	5.3	3.6	1.9	.362	.690	.352	.240	.102

Sources: U.S. data: U.S. Department of Labor, Bureau of Labor Statistics unpublished data. Japanese data: Japan Industrial Safety and Health Association 1985, 1986, 1988b.

[a]Injury and illness incidence rates per hundred full-time workers: cases involving days away from work; cases involving restricted work activity only have been excluded.

[b]Injury and illness incidence rates per 200,000 hours worked. This is the same as the hundred full-time worker basis. Injury and illness rates normally reported on a 1 million hours worked basis have been converted to a 200,000 hours basis for the sake of comparison.

the manufacturing industry provide evidence that there are, in fact, substantial differences in the work injury and illness experiences of the two countries. I will explore these differences further by reviewing U.S. and Japanese experiences at the corporate and plant levels. Before this analysis, however, I will first consider the relationship between establishment size and injury and illness experience in the United States and Japan.

Establishment Size and Occupational Injuries and Illnesses

Table 6-5 provides work injury and illness data for the manufacturing industries of the United States and Japan by establishment size. Comparisons are difficult because the establishment size categories are not exactly the same. Nevertheless, some interesting differences emerge.

Looking first at the U.S. data, we see that for 1984, 1985, and 1987 work injury and illness rates decline as establishment size increases. In 1987, the rate for establishments with 100 to 499 employees was 5.4 lost-time injuries and illnesses per hundred full-time workers or 2.3 times as high as the rate for establishments with 1,000 or more employees. Furthermore, the rate for establishments with 500 to 999 employees was 3.7 or 1.6 times as high as the rate for establishments with 1,000 or more employees. The disparities are only somewhat greater in 1984 and 1985.

The disparity is much greater in Japan. The 1987 rate for establishments with 100 to 299 employees was 7.5 times as great as the rate for establishments with 1,000 or more employees. Rates for establishments with 300 to 499 employees and with 500 to 999 employees were 3.8 and

2.6 times as great respectively as for establishments with 1,000 or more employees. The disparities are somewhat smaller in 1984 and 1985.

Since the categories are not identical, we cannot compare experiences directly. We can see, however, that in 1987 the rate for U.S. establishments with 1,000 or more employees was 39 percent lower than the average for all establishments with 100 or more employees. By comparison, for the same year, the rate for Japanese establishments with 1,000 or more employees was 75 percent lower than the average for all establishments with 100 or more employees.[7]

The U.S.-Japanese disparities were smaller in 1984 and 1985 but still substantial. In 1984, the work injury and illness rate for Japanese establishments with 1,000 or more employees was 72 percent lower than the rate for establishments with 100 or more employees, whereas it was 49 percent lower for comparable categories in the United States. In 1985, the comparable figures were 71 percent lower in Japan and 50 percent lower in the United States. Looked at from another perspective, smaller establishments in Japan have much poorer injury and illness rates relative to large establishments than in the United States.

These findings are consistent with my observations. Smaller establishments in Japan do not see the voluntary promotion of occupational safety and health as in their enlightened self-interest as master employers do. In the United States, however, OSHA (and state-plan OSHAs), labor unions, and even individuals are more aggressive in promoting adherence to minimal safety and health standards. This suggests that the punitive approach to safety and health regulation in the United States and adversarial labor-management relations with respect to safety and health issues may benefit employees in smaller establishments. The cooperative Japanese approach to occupational safety and health, which works so well for large establishments, may not work nearly as well for small establishments.

Company- and Plant-Level Comparisons

Given the enormous disparities between the United States and Japan in work injury and illness rates in the private sector, the manufacturing sector, and the automotive industry, it is instructive to consider differences at the firm and plant levels as well. I will do this by comparing the work injury and illness experiences of several Japanese automotive firms, including Jidosha, in Japan and in the United States. I will also compare their experiences to one another and the experiences of their U.S. operations to those of American automotive firms.

In making such comparisons, it is important to keep in mind the many sources of bias that distort comparisons of international work injury and illness experiences. The various conditions noted earlier in this chapter also affect comparisons of data at the company level. One additional factor is the time period being considered. The years 1987 and 1988 mark a radical change in reported work injury and illness rates for the U.S. auto industry because of the large fines imposed for underrecording during that time. Comparisons of rates from before that period to rates during or after it are therefore suspect. In addition, there are relatively great differences in how long the Japanese transplants have been in operation, which affects not only the availability of data but the level of experience of workers, and whether or not these plants are operating at full speed. Because occupational safety and health issues have been so controversial in labor-management relations at many transplants, the availability of data is also affected by the reluctance of managers to divulge specific information about their injury and illness rates. I am fortunate to have gained access to as much data as I have.

Table 6-6 repeats data from chapter 5 on injury and illness rates at Jidosha USA as well as information on how these rates compare with those for Jidosha in Japan. The table also contains data for two other major Japanese automotive firms, fictitiously named Kuruma and Norimono. Unlike Jidosha, neither Kuruma nor Norimono are unionized in the United States. Both have operated their U.S. plants for significantly longer than Jidosha has. Rates for the U.S. subsidiaries are included in table 6-6 as multiples of the rates for the parent firms.

As table 6-6 indicates, the lost-time rates for the U.S. subsidiaries of Jidosha, Kuruma, and Norimono are between 6 times and an astonishing 200 times as high as those for their Japanese parents. Jidosha accounted for the highest figure. Its lost-time injuries and illnesses occurred at a rate 17 times as high as its corporate parent in the only other year for which data are available. Kuruma USA had lost-time rates from 28 to 56 times as high as its parent; and Norimono USA had rates from 6 to 160 times as high.

No one transplant had clearly superior injury and illness experience. Jidosha's lost-time incidence rate of 1.0 in its first ten months of operation, in fiscal year 1987, was far lower than Norimono's rate of 3.1; in 1988, Jidosha's was much higher (12.4 versus 6.7); and in 1989, they were the same. In addition, Norimono's total incidence rate was much lower than the estimated total incidence rate at Jidosha during the same year. Figures for the latter period indicate the relative performance of these plants more accurately since Jidosha had a slow line speed during its first ten months. For the years 1985–87, Kuruma had a somewhat higher lost-time rate than Norimono (an average of 3.03 versus 2.33),

Table 6-6. Injury and Illness Experiences of U.S. Subsidiaries of Japanese Automotive Firms

	Jidosha USA			Kuruma USA			Norimono USA		
Year[a]	Total Incidence Rate[b]	Lost-Time Incidence Rate[c]	Multiple of Japanese Parent[d]	Total Incidence Rate[b]	Lost-Time Incidence Rate[c]	Multiple of Japanese Parent[d]	Total Incidence Rate[b]	Lost-Time Incidence Rate[c]	Multiple of Japanese Parent[d]
1989	32.2[e]	5.9[e]	N.A.	N.A.	N.A.	N.A.	13.9	5.9	N.A.
1988	44.4	12.4	200	N.A.	N.A.	N.A.	14.0	6.7	160
1987	N.A.	1.0[f]	17[f]	N.A.	3.5[g]	47[g]	8.9	3.1	86
1986	N.A.	N.A.	N.A.	N.A.	2.1	28	12.7	3.1	119
1985	N.A.	N.A.	N.A.	N.A.	3.5(est.)	56	9.1	.8	27
1984	N.A.	N.A.	N.A.	N.A.	N.A.	N.A.	6.7	.2	6

Sources: Internal company documents; Association of Japanese Safety and Health Managers of the Japanese Automobile Manufacturers' Association.

N.A. = not available.

[a]Some data are for fiscal years and some are for calendar years.

[b]Total recordable injuries and illnesses per 200,000 hours worked.

[c]Lost-time injury and illness cases per 200,000 hours worked. Jidosha USA and Norimono USA lost-time incidence rates do not include cases involving only restricted work activity. It is unclear whether such cases are included or excluded from Kuruma USA data.

[d]Lost-time rate of U.S. subsidiary stated as a multiple of lost-time rate of Japanese parent for closest comparable time period.

[e]First nine months only.

[f]First ten months of operation, spanning 1987 and 1988.

[g]Incidence rate data for U.S. subsidiary are for first six months only.

but Jidosha was not in operation at that time. More data are needed before we can draw any conclusions about the relative injury and illness experiences of these transplants. The data give us no reason to believe, however, that the rates for Jidosha USA were substantially out of line with those for other Japanese transplants.

The performance of Jidosha USA and other unionized Japanese transplants relative to that of nonunionized transplants is worthy of further study. The UAW has been attempting to organize the nonunionized transplants largely on the grounds that the union is necessary to protect the health and safety of the workers, which union organizers contend is compromised in management's relentless pursuit of higher productivity. While this argument has intuitive appeal, the data do not allow us to draw any conclusions.

Comparisons among Japanese Firms

Referring to figure 6-2, which contains lost-work incidence data for the Japanese operations of the ten leading Japanese automotive firms, we see that, of the three firms we are considering, Norimono had the lowest lost-work incidence rate for all four years. Norimono also had the lowest four-year average (.1675) of the ten firms. Jidosha had the highest average lost-work incidence rate of the three firms (.3950), ranking it sixth among the ten Japanese automotive firms. The lost-work incidence rates for Jidosha were above the industry average of .38075, whereas the rates for Norimono and Kuruma (.3375) were below it.

Figure 6-3 contains information on severity rates for these same ten Japanese firms. Norimono had the lowest four-year average among the three firms considered (.06575 lost workdays per thousand hours worked), ranking it fourth overall. Kuruma followed, with a rate of .08925, ranking it seventh. Jidosha was again last among the three, with a four-year average of .09475. This placed Jidosha tenth among the ten firms. (This low ranking is no doubt due to the two fatalities at Jidosha in 1987 and 1988.) Both Jidosha and Kuruma had higher work severity rates than the industry average, .0705; only Norimono had a lower rate of the three firms considered here.

Jidosha's slightly below-average lost-work incidence rate and its substantially below-average work severity rate have implications for the U.S.-Japanese comparisons in this study. It suggests that had I studied another Japanese automotive firm, the disparities between the United States and Japan might have been even greater.

The most significant feature of these data may be that they are available. Their collection and dissemination by the Committee of Safety

Figure 6-2. Lost-Work Frequency Experiences: A Ranking of Ten Japanese Automotive Firms

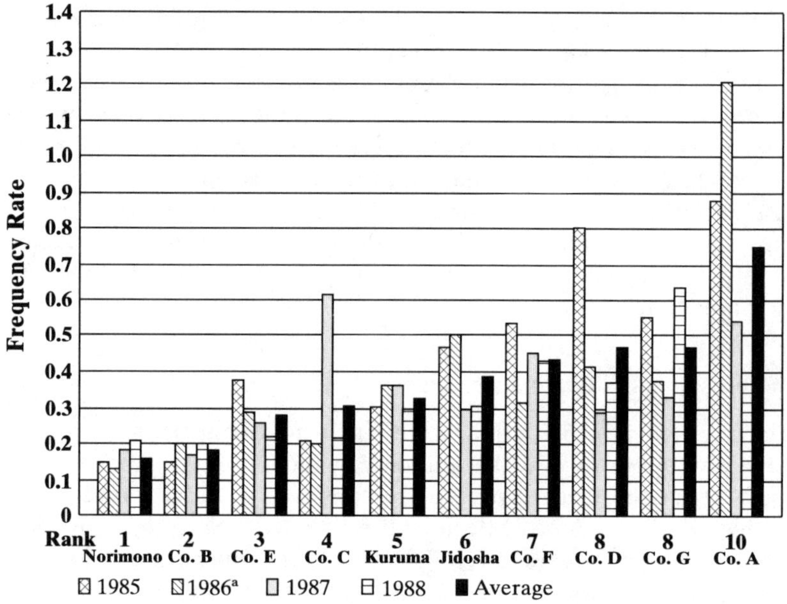

Source: Japanese Automobile Manufacturers' Association.
[a]1986 figures are for first ten months only.

and Health Managers of the Japanese Automobile Manufacturers' Association indicates there is far more cooperation among auto industry firms in Japan than in the United States. There is also competition, which has both positive and negative results. The competition undoubtedly spurs firms to work hard to reduce their incidence of injuries and illnesses, but it also encourages underrecording. Norimono's rates, in particular, are viewed with considerable skepticism among competitors in the Japanese automotive industry. In a rather blunt statement for a Japanese, a safety and health manager at Jidosha stated in reference to the statistics, "No one believes them." The American safety and health manager of Norimono USA was also very skeptical of their accuracy.

Operations of U.S. and Japanese Firms

One might expect that the ultimate test of the adaptability of Japanese safety and health management techniques to the U.S. work environment

Figure 6-3. Lost-Work Severity Experiences: A Ranking of Ten Japanese Automotive Firms

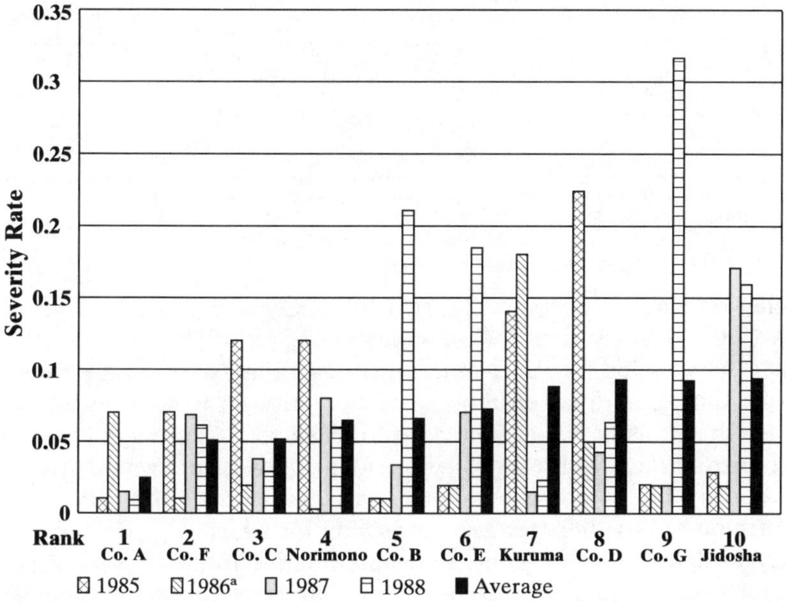

Source: Japanese Automobile Manufacturers' Association.
[a]1986 figures are for first ten months only.

would be the work injury and illness experiences of U.S. operations of Japanese transplants relative to those of traditional U.S. automotive firms. Unfortunately, the results are ambiguous. I will first consider the data and then the reasons for the ambiguity.

Table 6-7 presents work injury and illness data for several recent years for U.S. automotive plants employing more than twenty-five hundred employees. These are, for the most part, major assembly plants. Combined with data from table 6-6 and chapter 5, these data make it possible to draw some tentative conclusions regarding the incidence of reported work injuries and illnesses at Jidosha and at other Japanese automotive transplants compared with U.S. firms.[8]

These statistics show that the rates for the Japanese transplants are mixed relative to the industry averages. Kuruma USA had lower lost-workday injury and illness rates in the two most recent years for which data are available (3.5 versus 6.4 in 1987 and 2.1 versus 2.8 in 1986 but

Table 6-7. U.S. Motor Vehicle Industry and Illness Experiences of U.S. Plants Employing More Than 2,500 Employees[a]

	Lost-Workday Incidence Rate[b]	Total Incidence Rate[b]
1987	6.4	13.4
1986	2.8	6.2
1985	2.2	5.2
1984	2.5	6.0
1983	2.4	5.8

Source: UAW Health and Safety Department 1989, based on data supplied by the U.S. Department of Labor Bureau of Labor Statistics.

[a]Statistics are for motor vehicle and car body plants (SIC 3711).

[b]Injuries and illnesses per 200,000 hours worked.

a higher rate in 1985 (3.5 versus 2.2). The three-year average for Kuruma USA (3.03) is lower than for the industry (3.8).

As I have noted, the lost-time injury and illness frequency rate at Jidosha during its first ten months of operation (1.0) was lower than that for the industry, which increased from 2.8 to 6.4 during this period. More current rates for the industry are not available, but I would imagine that Jidosha's rates would be higher than those for the industry.[9]

Norimono USA, the Japanese transplant for which we have the most complete data, also shows mixed results relative to industry averages. Its total incidence rate is somewhat worse, but its lost workday frequency rate is somewhat better. For the period 1984–87, Norimono USA had a higher-than-average total injury and illness frequency rate in three years and a lower-than-average rate in one year. Its average total injury and illness frequency rate for these years (9.35) was 21 percent higher than the average for all U.S. automotive plants employing more than twenty-five hundred workers, which had a rate of 7.7. In contrast, in the same four years, Norimono USA had a lower-than-average lost-work injury and illness frequency rate in three years and a higher rate in one year. Norimono's four-year average (1.8 lost-work injuries and illnesses per 200,000 hours worked) was 48 percent lower than the industry average (3.48).

Observations on the Data

The relative experiences of American and Japanese firms obviously require continuing research. Based on the quantitative and qualitative data that are available, the safety and health programs and conditions of major Japanese firms in the United States and Japan deserve mixed reviews. They are superior in many ways, yet in other ways, they do not compare favorably. The safety and health operations of Japanese auto firms are superior in Japan to those in the United States but not to the degree that the data suggest at first glance. The superiority of the Japanese operations is attributable, for the most part, to the greater degree of worker compliance with established work rules. The work injury and illness data for Japan do not accurately portray the differences in safety and health conditions because of the apparently massive underreporting of work injuries and illnesses.

Japanese production techniques require that managers and workers exert greater control over the work environment and work practices than their American counterparts to ensure the efficiency and quality of production. As a result, line managers and individual workers in Japan are delegated far greater responsibility for safety and health.

In general, Japanese auto firms in both Japan and the United States are cleaner and tidier than American-owned auto firms. The workers, especially those in Japan, pay greater attention to safety and health rules and follow work practices more carefully. There are thus fewer traditional injuries resulting from carelessness, poor housekeeping, or the disregard of rules.

At the same time, the amount of stress, both physical and psychological, imposed on workers at transplants seems to be a problem. Efforts to use 100 percent of each worker's time in efficient, precisely defined routines appear to be taking their toll, most evidently in the number of workers with cumulative trauma disorders. The experience of Japanese transplants seems to suggest that a high incidence of CTDs is an expected outcome of the use of the Japanese manufacturing system in the United States. This is a disturbing trend that merits watching.

In addition, it appears that the protection from workplace hazards in small firms in Japan is inferior to that provided in small firms in the United States. The greater degree of disparity in safety and health conditions in small versus large firms in Japan relative to the United States reflects the greater reliance on voluntary employer efforts to promote safety and health in Japan. It also reflects the greater emphasis on individual rights in the United States.

The differences in the strengths and weaknesses of the U.S. and Japanese styles of worker protection have created the right conditions for the mutually beneficial transfer of safety and health policies and practices. Moreover, the experiences of U.S. subsidiaries of Japanese automotive firms suggest that the cross-national adaptation of approaches is indeed feasible. I discuss this potential further in the next and final chapter.

7

Conclusions and Recommendations

As we have seen, the regulation and management of occupational safety and health in Japan are in many ways consistent with more general Japanese regulatory and managerial practices. Japan's success in improving safety and health in the workplace derives from many of the same social, economic, political, and legal characteristics that contributed to its miraculous postwar economic recovery. It is not surprising, therefore, that there are many differences between the Japanese and the American approach to the regulation and management of occupational safety and health. These differences create opportunities for mutual learning, as well as the potential for cross-national adaptation of some practices.

At the beginning of this book I cited two myths about occupational safety and health conditions in Japan: the myth that Japanese workers are routinely and severely abused and the myth that the Japanese workplace is a paradise. By now it should be clear that neither of these myths captures the full complexity of conditions in Japan, or in Japanese transplants, yet both have elements of truth in them.

The myth of the abused worker suggests that the Japanese have built their economic success on the backs of their workers, who are overworked, underpaid, and forced to labor under hazardous conditions that we in "more civilized" countries would not tolerate. It is in this context that we occasionally hear international pleas to increase their wages, reduce their working hours, and improve their working conditions. Ostensibly, these pleas arise out of concern, but many critics hope that such measures will impose greater cost burdens on Japanese producers (or, more euphemistically, eliminate their "unfair advantages").

A close look at the actual conditions in Japanese industry indicates that the situation is much more complicated than the myth suggests. It is true that Japanese workers work significantly longer hours, on average, than do American and Western European workers, but discontent with the situation is resulting in mounting pressure, particularly from labor unions and younger workers, to reduce the workweek.

It is also true that Japanese wages in the past were low relative to those of workers in the United States. By the early 1990s, however, hourly labor costs (including benefits and bonuses) in Japan probably exceeded those in the United States for major employers.[1] A better place to look for an explanation of U.S.-Japanese cost disparities is executive compensation. Top U.S. managers, particularly in the auto industry, are paid much more than their Japanese counterparts. Further, the disparity between the compensation of production workers and that of top-level executives is far greater than in Japan.

It is also true that in the early postwar period occupational safety and health conditions were very bad throughout Japanese industry. Moreover, in certain sectors of the economy—primarily among small subcontractors—poor working conditions are still the norm. Nonetheless, in spite of several significant weaknesses, the safety and health conditions and practices in major Japanese corporations are superior to those in comparable American firms in many important respects. Especially noteworthy is the assumption of responsibility for safety and health by managers and workers throughout the organization. Moreover, personnel in these firms see safety and health management as integral parts of the production system and seek to derive competitive advantages through minimizing accidents, injuries, and illnesses.

Although they are in some ways discouraging, the competitive implications for U.S. industry of the findings of this research are also encouraging. That Japan's economic success was not produced at the expense of the safety and health of workers (at least not of those in master employer firms) puts to rest yet another excuse for the relatively poor U.S. industrial and economic performance vis-à-vis the Japanese in recent years. The research also suggests that safety and health, product quality, and productivity can be positively related in practice in much the way safety and health rhetoric in the United States claims they are. The experiences of Japanese auto firms operating in the United States suggest that it is possible, using Japanese safety and health management techniques, to achieve significant improvements on many dimensions of workplace safety and health (as well as product quality and productivity) in the U.S. work environment.

I do not mean to suggest, however, that the United States should attempt to emulate the Japanese occupational safety and health system in its entirety, even if we could. There are aspects of the system that are culture-specific and not likely to be transferable to the United States.

Other elements would be rejected on the grounds that they violate basic American values regarding respect for individuals and individuality. And, in some cases, practices are clearly inferior to the American approach.

The discussion above of the myth of the abused worker also provides both supporting and contradictory evidence regarding the myth that the Japanese workplace is a paradise. In addition to the many favorable aspects of the management and regulation of worker safety and health in Japan, there is—to use a term from the *Star Wars* movies—a dark side to them as well. No doubt the relentless pursuit of ever-greater productivity has negative safety and health implications. Even if accidents are avoided, the work pressure inevitably takes its toll on workers. Responses to anonymous surveys indicate that high percentages of Japanese workers are unhappy with the work pace and feel overworked and exhausted (Japan Institute of Labor 1989; Solo 1990). The high rates of cumulative trauma disorders and the intensity of complaints about the pace of work in American transplants of Japanese firms underscore the seriousness of the problems.

Another aspect of the dark side of occupational safety and health in Japan is the disparate treatment of workers. Permanent employees of master employers, who comprise about one-third of the work force, are a privileged class in Japanese society. The other two-thirds are treated far less well. These include workers in smaller subsidiaries, part-time and temporary workers in the master employer firms, and virtually all women, who traditionally have been considered temporary workers. This is true with respect to wages, hours, and working conditions in general and with respect to occupational safety and health conditions in particular.[2]

Other aspects of the dark side of the occupational safety and health system pertain to the sublimation of the individual in the pursuit of group goals. The use of shame and peer group pressure, though highly effective as a form of discipline and deterrence, can have devastating effects on workers.

Many of these negative features of the Japanese system conflict with fundamental American values. Although Americans also make within-group/out-of-group distinctions, we are less comfortable with differential treatment based on these distinctions. The stifling of dissent, restrictions on individual freedom, and the extreme value placed on organizational loyalty are difficult for Americans to accept. It is hard to imagine a UAW worker at Ford, General Motors, or a union-organized Japanese transplant obediently wearing a yellow shirt or other badge of shame as punishment for violating a work safety rule. It is also hard to imagine workers hiding major injuries or claiming they occurred at home to avoid bringing shame on themselves and their work group. In fact, just the opposite occurs. American workers claim at-home injuries took place at work so they can get better benefits.

The disparity in safety and health conditions between Japanese master employers and their small subcontractors and Third World subsidiaries raises a question that is seemingly puzzling. If occupational safety and health activities are, in fact, "profitable" in master employer firms, why do similar conditions not exist in the small subcontractors and Third World subsidiaries? The answer should now be obvious. First, the relationships between safety and health, productivity, and product quality are not as clearcut as Japanese managers and safety and health specialists would have one believe. The slogans proclaiming these relationships and the assertions of managers that the slogans are true must be taken as *tatemae*.

Second, managers in small subcontractors and in Third World subsidiaries of Japanese firms are not acting in their firms' enlightened self-interest. Because of poor management, the firms are thus forgoing competitive advantages that they could derive from improved occupational safety and health performance.

Third, the relationship of safety and health to productivity and product quality may be different in smaller Japanese firms and in subsidiaries. Accidents, injuries, and illnesses are probably less costly in these organizations since they probably have less invested in their workers in the form of training (this is also the case with temporary and part-time workers in major employers) and are paying less injury and illness compensation. Further, because their levels of capital investment are lower, equipment damage in the event of an accident is less costly.

Fourth, and finally, the manufacturing systems in these firms probably are not operating as close to peak efficiency and therefore probably suffer fewer operational losses when there is an accident.

The disparity in safety and health conditions between large and small firms can also be explained by the relationship between master employer and subcontractor firms. Master employer firms contract out many of their most hazardous jobs. They can also place relentless pressure on the subcontractors to cut costs. In addition, although master employer firms usually assist their principal subcontractors in the management of their safety and health activities, this assistance typically extends only to the first layer of subcontractors; lower-level subcontractors and Third World subsidiaries typically are ignored.

Positive and negative stereotypes are also common with respect to the transplants. Training, at least in the early stages of production, and control over traditional types of accidents and injuries does appear to be outstanding. At the same time, negative features of the system are also evident. In particular, there are indications that in many plants the Americans are viewed as outsiders. They are excluded from participating in key decisions, glass ceilings limit their advancement to top management, and there is a high rate of turnover. Most disturbing from a safety and health perspective, there is a high incidence of cumulative trauma disorders.

There are relatively few nondemanding jobs in which to place workers with CTDs. What will become of them? The evidence seems to suggest that, to a certain extent, they are viewed as expendable; they can be replaced by younger, healthier workers until they too wear out.

A Contingency Theory of Cooperation

The mixed findings of this study suggest a contingency theory regarding the benefits of labor-management-government cooperation in occupational safety and health. Basically, the hypothesis is that cooperation among these parties will have highly beneficial results if and only if each of these parties is assertive and has a minimum amount of power. This study attests to the effectiveness of cooperative safety and health relations among labor, management, and government in Japanese automotive firms both in Japan and, somewhat less clearly, in the United States. The abilities of these parties to work together to find cost-effective and mutually agreeable solutions to safety and health problems is exemplary.

Cooperative relations appear to work less well, however, when one or more of the parties are relatively powerless and unable to assert their rights. Small subcontractors, where workers are reluctant to complain to government regulators for fear of being found out and punished, are an obvious example. The situation is exacerbated by the nonpunitive nature of enforcement by the Labor Standards Bureau and by its tendency to focus on larger employers. Similar conditions of powerlessness are likely to exist in Third World countries, where workplace safety and health standards and their enforcement are much more lax (Wokutch 1990a). Likewise, the underreporting of injuries and illnesses by managers in major employers in Japan occurs as a result of the minimal risk of being caught and punished. This contrasts with the situation in the United States, where OSHA has put managers on notice that there are considerable financial risks associated with underreporting. Finally, workers in Japan may hide injuries and illnesses because of their or their unions' lack of power or assertiveness. Having been convinced by management that accidents are their fault or the fault of their supervisors, injured workers may hide their injuries out of a sense of shame or fear of retribution. An American worker, even in a Japanese-owned firm, is much more likely to blame management for accidents (e.g., an equipment defect) even if it was the worker's fault.

Referring back to figure 1-2, it would seem that the ideal environment for promoting occupational safety and health is one in which there are high levels of both cooperativeness and assertiveness. Research involving several other countries led me to the same conclusion (Wokutch 1990a).

If indeed this conclusion is correct, then both Japan and the United States have something to learn from each other. In the United States, labor, management, and the government need to work together more cooperatively. In Japan, labor and the government need to be more assertive, particularly with respect to smaller firms.

Japanese firms operating in the United States seem to provide an ideal environment for combining the best of Japanese and American approaches. These firms have been largely successful in replicating the cooperative Japanese corporate culture. Yet, because managers have to contend with workers and safety and health regulators who by nature are more assertive than those in Japan, they can avoid the negative consequences of unrestrained cooperation. Further, both workers and managers accord higher priority to safety and health than in U.S. firms. To date, however, observations of transplants provide only limited support for this assessment.

Meanwhile, the performance of the transplants in the areas of safety and health seem disappointing compared with the performance of these firms in Japan. There are several reasons for the disparities. First, the workers and managers are relatively inexperienced. Second, injury and illness recording is apparently more accurate at the transplants than at the parent firms. Third, U.S. workers resist following safe work practices to a much greater extent than do Japanese workers. Compliance is far better in U.S. subsidiaries of Japanese firms than in domestic auto firms, but it is nowhere near as good as in Japan. Compliance could be improved, but it is highly unlikely that the extent of compliance evident in Japan will ever be achieved in U.S. plants, unless, of course, a Japanese work force is imported. The effect of culture is just too great. Fourth, the Japanese manufacturing system may not suit U.S. workers. The unending pursuit of increased productivity and efforts to use 100 percent of a worker's time have resulted in a high incidence of cumulative trauma disorders. These conditions, which have led to higher reported work injury and illness rates in the transplants, have apparently outweighed the benefits of the greater assertiveness of labor and regulators. It is unlikely that injury and illness rates at any U.S. automotive operations will ever match reported rates for Japanese operations, but there are ample opportunities for improvement.

Lessons America Can Learn from Japan

The experience of Japanese auto firms operating in the United States demonstrates that it is indeed possible to adapt certain positive features of the Japanese style of worker protection to the U.S. work environment. One such feature is the spirit of cooperation among labor, management, and government regulators regarding safety and health. This cooperation has come about as a result of the commitment by top management to occupational safety and health and a recognition throughout these organizations that safety and health management is an integral part of the production system and everyone's responsibility. The delineation of specific safety and health responsibilities for workers and managers at all levels of the plant is a useful first step toward developing such a commitment.

The cynicism and bad habits that have developed over time in U.S. plants regarding occupational safety and health make it questionable whether a turnaround in attitudes and practices can be achieved. Even the Japanese transplants have to fight to keep old habits from reemerging. Certainly a new venture such as General Motors' Saturn plant with its new workers, new contract, and apparently new work attitudes would make a better candidate than an existing U.S. plant for the development of this cooperative spirit.

The initiative clearly belongs to management if relations with safety and health regulators are to be restructured. The U.S. subsidiaries of Japanese auto firms have developed relatively congenial relations with the same government safety and health regulators about whom their counterparts at U.S. firms complain bitterly, and they developed these relations while their plants were under intense scrutiny as highly publicized new employers. Given the considerable differences in the nature of the regulatory environment in the United States and Japan, this is impressive. OSHA's ill-fated cooperative programs (Praise, Star, and Try) under the Reagan administration met with less than lukewarm reception from industry (Wokutch 1990a). The ball is now in industry's court. At the same time regulators must not be so taken in by programs at the Japanese subsidiaries that they become co-opted. Clearly, the high degree of assertiveness of safety and health regulators in the United

States is a strength that should not be abandoned in the pursuit of greater cooperation.

Lessons Japan Can Learn from America

It may be surprising to Americans accustomed to criticism in the face of Japanese industrial and economic successes that in some ways the American style of worker protection is clearly better. First, the regulatory system provides the worker as an individual with greater protection. The great disparity in protection afforded Japanese workers is not inherent in the system. Greater assertiveness by government safety and health regulators is possible and would be highly beneficial, especially to workers in smaller firms.

Second, the traditional American manufacturing system is more forgiving than the Japanese system of individual differences in physical characteristics as well as preferences about how to perform a job. While the Japanese approach appears to work well in Japan, it is inappropriate in the U.S. work environment. Managers of Japanese transplants would be well advised to make greater accommodations for the individual differences among their workers.

Third, because they are more assertive, American unions offer their members greater protection than do the cooperative enterprise unions in Japan. By being more assertive, Japanese unions could substantially improve the treatment of workers, especially part-time and temporary employees.

Other Lessons

Clearly, comparing work injury and illness statistics for Japan and the United States is very difficult. Nevertheless, it is essential that Japanese companies operating in the United States keep statistics that would make such comparisons possible. The absence of standard company-wide definitions is a particular problem. The promulgation of such definitions would be time-consuming, requiring, in many places, recordkeeping to comply with both company definitions and government definitions. This practice would not eliminate all the bias involved in comparisons, but it would be a major step toward establishing better control over work injuries and illnesses. It would also send a message that safety and health really are as important as quality, productivity, and cost control, all of which are evaluated using company-wide measures.

Experiences of the last few years, including, most notably, Union Carbide's tragedy in Bhopal, India, suggest that far more attention needs to be given to workplace safety and health in the operations of multinational corporations in developing countries. Many aspects of the programs used to promote safety and health in U.S. plants of Japanese firms could be adapted with positive results to subsidiaries of U.S. and Japanese firms in these countries. The safety and health components of preproduction training programs and ongoing relationships with "sister factories" in the home country merit particular consideration.

Corporate Social Responsibility and Competitive Implications

In chapter 1, I raised two questions regarding the social responsibility of corporations in matters of occupational safety and health: (1) are firms doing everything they can with respect to occupational safety and health that is in their own enlightened self-interest (or long-run profit interests) and (2) are they undertaking some activities in this area for altruistic reasons? Obviously, it is difficult to generalize, but, at least with respect to most U.S. firms, the evidence suggests that the answer to the first question is a resounding no. Many American safety and health managers concede that they simply have not done a good enough job in demonstrating to top management the costs of accidents, injuries, and illnesses and the benefits to be achieved by reducing them. Perhaps they will have a more receptive audience as health-care costs escalate even further.

The answer with respect to Japanese firms is more ambiguous. As noted, one reason for the very poor safety and health performance of small firms in Japan is the apparent inability of their managers to recognize what is in their self-interest. Even larger employers have weaknesses that could be addressed to the financial benefit of the firm. Probably the greatest of these weaknesses is the underrecording of injuries and illnesses, especially those that go untreated. It is hard to imagine that an injured or sick employee could work at peak capacity. The Labor Standards Bureau and company unions need to take more assertive roles in forcing firms to do what is in their own best interest.

Whether some safety and health activities are undertaken for altruistic motives is also difficult, if not impossible, to determine. Evidence points to the overwhelming dominance, however, of enlightened self-interest as the motivation for Japanese occupational safety and health activities. The reported disparity in safety and health practices and conditions in Japanese-owned plants operating in Japan and the United States and those operating in developing countries supports this assertion. If respect for the intrinsic value of human life were included in calculations of the financial costs of injuries and illnesses, somewhat more effort would probably be made to extend company safety and health practices to operations in less developed countries.

There clearly are major differences in the management of corporate social responsibility in Japanese and U.S. firms. Unlike so many other areas for comparison, however, the benefits are not all one-sided. Occupational safety and health by major Japanese automotive firms is certainly one dimension of CSR in which the Japanese have a competitive advantage over the United States. The overall Japanese approach to CSR is not appropriate in the U.S. work environment, however, given the far greater importance placed on individual rights codified in law and given that the social environment is more dynamic and challenging than in Japan. In light of these differences, Japanese firms have done surprisingly well in adapting to this more dynamic, individualistic, and conflict-oriented society. At the same time, as I discussed in chapter 2, many Japanese firms have had serious problems in adapting to this environment with respect to sexual and racial discrimination, product liability, business ethics, and even philanthropic contributions. The treatment of workers in Japan, South Africa, and developing countries that is not affected by American law and public opinion deviate even more from established U.S. social norms.

In spite of their problems, Japanese companies project remarkably favorable public images in the United States. They do so by paying a great deal of attention to how they are perceived. In fact, many major Japanese corporations are more concerned about their image in the United States than in Japan. The size of the market is obviously important, but so is the precariousness of their presence. Japanese firms, industry associations, and the Japanese government have been very effective in positively influencing public opinion in the United States (*Business Week* 1988; Choate 1990; Prestowitz 1989). Their activities have included funding public television programs about Japan, corporate philanthropy, traditional public relations, and lobbying in Washington with the best public relations agents money can buy. They have also funded research at major universities and think tanks that they believe will produce results that will support their economic and political objectives. In many cases, they have hired high-priced public relations firms and put them in charge of managing their corporate image. But there is substance behind the image. At least in the auto industry, Japanese firms have been able to avoid negative publicity by paying meticulous attention to details in the planning and implementation of their ventures in the United States.

For most Japanese firms, the debate about the social desirability of their activities has focused primarily on issues of trade. Thus, much of the discussion has centered on theoretical issues concerning the benefits and costs of free trade versus protectionism. As the number of Japanese companies operating in the United States and the number of people directly affected by their activities continue to grow, other aspects of corporate social responsibility will take on greater importance, including affirmative action/discrimination, pollution, product liability, employee rights and responsibilities, antitrust issues, relations with South Africa, and business ethics.

The *tatemae/honne* distinction is useful for explaining many of the activities and public relations communications of Japanese companies in the United States. One needs to look closely at the behavior not only in Japan but in other countries, rather than just listen to the rhetoric.

The divergence between the reality and the image is a potential weakness of Japanese companies that U.S. companies could exploit to competitive advantage. The public is concerned about racial and sexual discrimination, employee rights in the work environment, operations of companies in less developed countries and South Africa, product and workplace safety, and other aspects of corporate social performance. U.S. companies outperform their Japanese competitors on many of these dimensions. Providing performance information in these areas that would allow others to make comparisons could be potentially beneficial. Other comparisons could be made directly (e.g., product safety test results) and could be equally favorable to U.S. firms. U.S. firms might also consider using some of the strategies employed by the Japanese in the areas of funded research and television programs, public relations, and lobbying.

U.S. firms also have operational advantages in the area of CSR management. The extent to which U.S. firms do not discriminate in hiring and promotion decisions, for example, affords them a larger pool of available talent from which to choose. The disparity on this dimension and the resultant competitive advantages are greatest for U.S. firms operating in Japan. A large pool of highly talented Japanese women is available to U.S. firms trying to penetrate the traditionally closed Japanese economy.

One should not underestimate Japanese competition even in the arena of corporate social performance. Evidence suggests that Japanese firms operating in the United States have adapted quickly to U.S. norms in their actions if not in their beliefs. Moreover, given the increasing importance of direct Japanese investment in the United States, improvements in the social performance of Japanese firms will have significant social benefits here. It is also possible that forced changes in behavior (for example, in the area of equal treatment of women and minorities) will bring about changes in underlying beliefs and values, at least in those Japanese working in the United States and possibly in other Japanese as well.

The United States and Japan have much to learn from each other regarding labor-management relations, the government regulation of industry, and the management of corporate social responsibility. The key to successful exchanges in these areas is to find ways to promote the efficiency and cooperative relations associated with Japanese practices while maintaining the freedoms and protections for the individual so important in the United States. This will no doubt be difficult to do, but given the increasing interdependence of the U.S. and Japanese economies, it is essential that we try.

Notes

1. Introduction

1. I am indebted to David Vogel (1986) for this "national-style" terminology.

2. Womack, Jones, and Roos (1990) found that the Japanese transplants trail Japanese plants in Japan by about 25 percent in productivity. They attributed the lower figure primarily to the recent start of production of the transplant ventures. They found levels of quality to be about the same.

3. Much of the discussion of corporate social responsibility contained in this and subsequent chapters was previously published as an article entitled "Corporate Social Responsibility Japanese Style" (Wokutch 1990b).

4. Mafune 1988, Sethi 1975, and Sethi, Namiki, and Swanson 1984, as well as Wokutch 1990b, deal explicitly with this issue.

5. This economic or financial view of altruism is distinguished from the more traditional notion of altruism to avoid the debate about whether receiving certain psychic benefits from helping others disqualifies the act of helping from being altruistic. Advocates of a controversial school of ethics called psychological egoism contend that individuals undertake only those activities that are in their perceived self-interest (see Beauchamp and Bowie 1983 and Bowie 1991 for discussions of psychological egoism). Thus, psychological egoists would argue that no one can act altruistically. Even heroic acts such as sacrificing one's life to save others are viewed as being motivated by perceived self-interest. For the view of altruism discussed here to be shown to exist, one just needs to show that one does not expect to benefit *financially*.

6. This study is an extension of my earlier study. This study provides far more in-depth analysis of the intriguing differences in the regulation and management of safety and health in the United States and Japan than was possible in the earlier broader-based book.

2. The Environmental Context for Occupational Safety and Health Activities in Japan

1. David Halberstam (1986) provides a fascinating account of the intense and pivotal labor-management conflict at Nissan in the early postwar period which management eventually won.

2. These suicides were the subject of considerable debate. Government officials contended that many were unrelated to privatization and that the number was normal.

3. Emperor Hirohito's reign was named Showa or "enlightened peace."

4. See William Bohnaker (1990) for an amusing discussion of the differences in Japanese behavior toward the environment in public versus private places.

5. For a more detailed discussion of environmental issues in Japan, see Beck (1986), McKean (1981), Pharr and Badaracco (1986), and Sethi (1975).

6. See Drucker (1981a, 1981b) for discussions of Confucian ethics as they apply to business in Japan.

3. The Safety and Health System in Japan

1. By 1990, the government's share of the budget had risen to 46 percent (Japan Industrial Safety and Health Association 1989).

2. A fuller discussion of cooperation and conflict among labor, management, and government in the United States is in Wokutch 1990a.

3. The OSHA recordkeeping change in 1983 affects this analysis and the analysis of several other dimensions of U.S. and Japanese operations reported in this study. For this reason, the differences in means and variances were considered for the periods 1974–82 and 1983–87 separately. For the period 1974–82, the analysis revealed that the mean number of field staff in Japan was significantly higher than the number of personnel in the United States ($t = 8.9973$, d.f. = 16.0, $p = .0000$). No significant difference in variances at the .05 significance level was found ($F = 4.01$, d.f. = 8/8, $p = .0664$).

For the years 1983–87, the mean number of field staff in Japan was significantly higher than the total number of OSHA personnel in the United States ($t = 70.3839$, approximate d.f. = 4.7, $p = .0001$), and the variance of U.S. personnel was significantly higher than the variance in the numbers of field staff ($F = 10.65$, d.f. = 4/4, $p = .0417$).

The statistical analyses of staffing levels and activities for the constituent time periods reported here and below should be viewed with caution, however, because the time periods are so short and consequently the sample sizes very small.

4. Looking at the 1974–82 and 1983–87 periods separately, the mean for Japan is significantly higher ($t = 18.5068$, d.f. = 16, $p = .0000$) but the variances in the former period are equal ($F = 3.48$, d.f. = 8/8, $p = .0970$). In the latter period, the mean for Japan is again significantly higher ($t = 167.1637$, d.f. = 8, $p = .0000$) and the variances again are equal ($F = 4.68$, d.f. = 4/4, $p = .1640$).

5. The averages reported here for the number of inspections per inspector and the number of inspections per field staff member in Japan are different from those reported in table 3–2 because data for 1972 and 1973 are not included in the averages reported here. They were dropped because U.S. data for those years were not available.

6. Again, because of the recordkeeping change in 1983, the means and

variances for 1974–82 and 1983–87 were considered separately. I look first at the data on inspections per inspector for Japan and then at the data on inspections per field staff member. For the 1974–82 time period, the mean number of inspections per inspector was not significantly different for Japan than for the United States (t = 2.1310, approximate d.f. = 9.4, p = .0608). For 1983–87, the mean was significantly higher for Japan (t = 2.6883, d.f. = 8, p = .0276). There was a significant difference in the variances for the period 1974–82, with the U.S. variance being higher (F = 11.57, d.f. = 8/8, p = .0023). For 1983–87, there was no significant difference (F = 6.51, d.f. = 4/4, p = .0969).

Comparisons of the number of inspections per U.S. inspector and inspections per Japanese field staff member also revealed that there were differences in the results for the overall time period and for the constituent periods. There was no difference in the means for the 1974–82 period (t = .7427, approximate d.f. = 9.3, p = .4761), but there was a significant difference for 1983–87 (t = 3.4489, d.f. = 8, p = .0087). The United States had a higher mean in the latter period. There was a significantly higher variance for the U.S. data for the 1974–82 period (F = 12.22, d.f. = 8/8, p = .0019), but the p-value for the 1983–87 period was slightly higher than the .05 significance criterion (F = 8.98, d.f. = 4/4, p = .0562).

7. Statistics on numbers of penalties were not available for either country.

4. Occupational Safety and Health at Jidosha

1. Much of this chapter and the next are based on interviews with Jidosha officials from various departments and on observations of work practices and the work environment over an extended period of time. Many points made here emerged as a result of interviews with groups of workers or managers and are not attributable to particular individuals. The pseudonym Jidosha has been inserted into direct quotes to maintain confidentiality and enhance readability.

2. These remarkable similarities were no doubt due in part to the use of group interviews, which were prepared for me at every company I visited. There were undoubtedly subtle pressures on the interviewees to give *tatemae* or socially acceptable answers. A cynic would suspect that the group interviews were set up so I would hear such answers, but it may also be that the Japanese are more comfortable with this format. On several occasions, the translator or others accompanying me noted that the interviewees were nervous because this was the first time they had been interviewed by a Westerner. Western researchers had obviously visited these companies before, but apparently none had spoken with safety and health personnel.

3. Jidosha officials use this figure in calculating a "presenteeism" rate of 99.7 percent (100 percent − .3 percent), which they prefer to discuss.

4. One American resident of Japan noted in the Readers Report section of *Business Week* (Littleboy 1989) that Japanese workers often forgo using their vacation days out of fear of losing bonus payments. I uncovered no evidence at Jidosha or other Japanese auto firms that bonus payments are linked to a

willingness to forgo vacations, but such a practice would not be surprising. Jidosha officials conceded that bonuses could be withheld if someone had excessive nonvacation absences.

5. Workers and managers tended to focus on safety issues, but most of their comments apply to health issues as well. Officials at Jidosha believe work-related illnesses are not a problem. This belief is probably erroneous and due to the underreporting of work-related illnesses.

6. In 1986, the maximum workers' compensation settlement for a work injury in Japan, fatal or otherwise, was approximately 100,000,000 yen or $667,000.

7. Since this was written there has been a significant reduction in the number of people working in the Safety and Work Environment Division. Nevertheless, staffing levels remain quite high compared with levels in U.S. companies.

8. This abbreviation comes from the Japanese phrase "*Sokuji Kiken Yochi*" or "on the spot danger prediction" used by Tokyo Kogyo Co., Ltd., in its zero-accident campaign. It is also explained that "SKY happens to be the same as sky in English, it sounds nice and gives the pleasant feeling" (Tanabe 1984: 97).

9. In discussing the popularity of smoking, an American visitor noted that several years ago his seat on a domestic flight in Japan was the only one designated nonsmoking; everyone around him smoked. Since then, smoking has been banned on domestic flights of less than one hour.

10. News accounts confirm that there is little use of recreational drugs on the job in Japan. There is more concern about the use of stimulants by tired and weary workers who take them in the belief that they will improve their performance or enable them to continue working beyond the point at which they might otherwise have to stop.

11. In comments on a draft of the manuscript for this book, T. J. Pempel (1990) of Cornell University pointed out what becomes obvious after several bouts of drinking with Japanese acquaintances, that the Japanese seem to get intoxicated with less alcohol, even controlling for physical size, than do Americans. He also observed that their work on the day following a drinking bout does not seem to be affected: "Hangovers, etc., seem miraculously to disappear." Whether this is because of a difference in physiology, as he speculates, or because workers put up a good front is uncertain. In either event, the minimal evidence of aftereffects explains in large measure the downplaying of the problem.

12. When comparing fiscal years with calendar years, the year in which the fiscal year *ends* is matched with the corresponding calendar year. Thus, the 1967–68 fiscal year is matched with calendar year 1968. For the years 1987 and 1988, it is unclear whether the Jidosha data and the industry data are for fiscal years or calendar years. This should not affect the analysis appreciably, however.

13. It would be acceptable practice in the United States to send an injured or ill worker to rehabilitation therapy during working hours, and, in fact, Jidosha USA has recently started doing this, but the original injury or illness would still need to be recorded.

14. On reading a draft of the manuscript for this book, public relations officials at Jidosha asked that this figure be "corrected" to 35°C (95°F) but provided no further explanation or evidence. Regardless of the actual temperatures, this *was* the worker's complaint.

15. In commenting later on a draft of the manuscript for this book, public relations officials contended that this disciplinary action had taken place more than ten years earlier and that employees are no longer disciplined in this manner. They also requested that reference to the incident be deleted.

5. Occupational Safety and Health at Jidosha USA

1. Subsequent reports placed Jidosha USA's total absenteeism at 10 percent. Total absenteeism, as the term implies, covers all absences, including military leave, vacation leave, sick leave, and so on. This rate is much higher than the rates reported for Jidosha in Japan and for other Japanese transplants but, according to Jidosha USA personnel, is still somewhat lower than at the Big Three's plants.

2. The patterns in turnover in U.S. and Japanese operations of Jidosha are very different. A large proportion of the turnover in Japan is accounted for by young women who quit to get married and raise a family or workers who quit to retire. In contrast, most of the turnover in the United States is for reasons directly related to the work itself. Subsequent reports placed the turnover rate at Jidosha USA at or above the rate for the U.S. auto industry, indicating a growing disenchantment with working conditions at Jidosha. Although he did not have exact figures, the chairman of the UAW local at the plant described the turnover in this period as "shocking." Even later reports indicated turnover was stabilizing at a lower level as an ever-increasing percentage of the work force consisted of workers who had adapted to the pace and conditions of work.

3. The following statement, entitled "Community with the Community," appears on a large plaque that hangs in the main passageway in the Jidosha USA plant: "Jidosha believes meeting our social responsibilities is one of the very important corporate activities. Since its founding Jidosha has worked hard to share benefits with local communities. At Jidosha we are more than just an employer. Jidosha is a corporate citizen: a part of the mosaic of the [local community]. There are more than altruistic reasons behind this belief. Practically speaking, a healthy vibrant community is vital to the future of Jidosha. Jidosha hopes that through activities such as the Jidosha-Youth for Understanding sponsorship and United Foundation participation, to become an enterprise known and trusted the world over."

4. Although Jidosha USA is partially self-insured for workers' compensation, it carries a policy for excess coverage. Because workers' compensation claims, especially for CTDs, have been so numerous, the insurance company has been required to pay for a great many claims.

5. During the UAW organizing campaign at Nissan in Smyrna, Tennessee, workers there also complained that supervisors discouraged them from drinking too much water, in this case so that they would not have to take breaks as often to urinate (Foust et al. 1989).

6. In traditional UAW-organized plants, foremen perform only supervisory functions and do not work on production.

7. The real magnitude of the United States–Japan CTD rate disparity is not ascertainable through the research design employed in this study. A UAW safety and health specialist suggested a research approach that, although very costly, would provide a clearer picture. He suggested arranging for an independent industrial physician to examine workers in both countries for evidence of CTDs. Of course, even this approach would entail a bias to the extent that the diagnosis would rely on the self-reporting of symptoms, which the Japanese would be far less likely to report.

8. Since these data were collected, there has been a decrease in the number of corporate safety and health staff at Jidosha in Japan and an increase in the number of workers at Jidosha USA. The difference in staffing levels, however, remains about the same.

9. The total injury and illness rate and lost-time injury and illness rate for Japan are based on per million hours worked, whereas the rates for the United States are based on per hundred full-time workers. One hundred workers are assumed to work 200,000 hours per year, based on the assumption of a forty-hour workweek and a fifty-week work year, so that one can divide the Japanese rates by a factor of five to compare them with the U.S. rates.

10. In Japan, charges are also added for amputations or other permanent total or partial disabilities. Because these charges vary depending on the degree of disability and precise data are not available, I was unable to factor out these charges as I was for fatalities.

11. These figures are for total CTDs, that is, those involving lost work combined with those involving no lost work.

12. Japanese managers gave these estimates to a Japanese researcher studying personnel and labor relations issues at Japanese auto transplants. He was not specifically studying safety and health, but the rates of CTDs were cited as a particularly troublesome problem in their U.S. operations.

13. Fiscal year data for Jidosha USA are compared with the most closely corresponding calendar year data for the United States.

14. Using U.S. Bureau of Labor Statistics terminology, this is the lost-workday case rate.

15. Statistics reported in this section for establishments employing one thousand or more workers are from unpublished 1990 data from the U.S. Department of Labor, Bureau of Labor Statistics.

16. These figures are for days away from work; they do not include restricted work activity days.

17. Data for 1987 and 1988 on rates of CTDs for auto industry establishments with one thousand or more employees are from 1990 unpublished statistics provided by the U.S. Department of Labor, Bureau of Labor Statistics.

6. Work Injury and Illness Experiences in the United States and Japan

1. According to the U.S. government publication *Occupational Injuries and Illnesses in the United States by Industry, 1987*, "*Lost workday cases* are cases which involve days away from work, or days of restricted work activity, or both. *Lost workday cases involving days away from work* are those cases which result in days away from work or a combination of days away from work and days of restricted work activity. *Lost workday cases involving restricted work activity* are those cases which result in restricted work activity only" (U.S. Department of Labor, Bureau of Labor Statistics 1990b: 76).

In the United States, companies sometimes refer to lost workday cases as lost-time incidents or lost-work incidents. In Japan, what the U.S. Bureau of Labor Statistics would call lost workday cases involving days away from work are referred to as lost-time cases, lost-work cases, or work-shutdown cases.

2. The aim of these campaigns is understood to be the achievement of zero illnesses as well as zero injuries.

3. There is confusion or lack of precision in the English-language translations of published Japanese work injury and illness statistics. According to the *Dictionary of Labor Terms* (Ministry of Labor 1973: 880), a work accident refers to "Injuries, illnesses and deaths which occur because of such things as buildings, equipment, materials, gas, steam, dust or workers' own misconduct at the workplace." In addition to this term, various publications used to compile tables 6-1 to 6-5 and figure 6-1 use the terms injuries, injuries and illnesses, and casualties. Based on extensive discussions with officials from the Ministry of Labor, I have concluded that these terms are used interchangeably to denote what we in the United States understand to be injuries and illnesses combined. Even if this assessment were incorrect, it would not make a great deal of difference in the following analysis since there are so few recorded work illnesses in Japan. In 1988, for example, only 12,523 occupational illnesses were recorded nationwide. In contrast, 240,700 were reported in the same year in the United States (U.S. Department of Labor, Bureau of Labor Statistics 1990b).

4. To be more precise, cases involving *only* restricted work activities have been excluded. Cases that involve both lost workdays and restricted work activity have been retained.

5. Heinrich did not address the issue of illnesses in his analysis, but because these comprise only a small percentage of total injuries and illnesses, their inclusion should not significantly affect the analysis that follows.

6. A more precise calculation would use just injury data rather than injury and illness data combined, but these are not available for Japan.

7. These comparisons understate the disparity between large and small establishments in both the United States and Japan in that I am comparing the experiences of large establishments with those of the industry, which includes large establishments. As we saw, greater disparities are evidenced when the experiences of large and small establishments are compared with each other rather than with industry averages.

8. Data in table 6-7 are actually for all plants, including Japanese-owned plants, employing more than twenty-five hundred workers. Nonetheless, the overwhelming majority of these plants are American-owned.

9. For confidentiality reasons, the U.S. Bureau of Labor Statistics will no longer provide data by industry for establishments with twenty-five hundred or more employees. The view that Jidosha USA had a worse injury and illness record than other large automotive establishments is corroborated, however, by data from that agency on industry establishments with one thousand or more employees. These establishments had a lost-workday incidence rate of 8.9 and a total incidence rate of 26.8 in 1988, considerably lower than the respective rates for Jidosha USA.

7. Conclusions and Recommendations

1. The U.S. Department of Labor, Bureau of Labor Statistics (1990a) placed average Japanese hourly compensation costs for all manufacturing firms at $12.63 in 1989 compared with $14.31 for the United States in the same year. It is likely, however, that compensation for employees in major Japanese firms exceeds the average compensation levels to a greater extent than in comparable firms in the United States.

2. I do not specifically address the occupational safety and health implications of the second-class corporate citizenship of women since women have been systematically excluded from production jobs in major Japanese auto firms. One result of the discrimination against women is that they may be disproportionately represented in hazardous production jobs in smaller firms. This issue deserves further study.

References

Abegglen, James C., and George Stalk, Jr., 1985. *Kaisha: The Japanese Corporation*. New York: Basic Books.

Altshuler, Alan, Martin Anderson, Daniel Jones, Daniel Roos, and James Womack. 1984. *The Future of the Automobile*. Cambridge, Mass.: MIT Press.

Arthur Andersen and Company. 1979. *Cost of Government Regulation Study of the Business Roundtable*. Chicago: A. Andersen.

Ashford, Nicholas A. 1976. *Crisis in the Workplace: Occupational Disease and Injury. A Report to the Ford Foundation*. Cambridge, Mass.: MIT Press.

Automotive News. 1990. *Market Data Book*.

Bacow, Lawrence S. 1980. *Bargaining for Job Safety and Health*. Cambridge, Mass.: MIT Press.

Badaracco, Joseph L., Jr., 1985. *Loading the Dice: A Five Country Study of Vinyl Chloride Regulation*. Boston: Harvard Business School Press.

Beauchamp, Tom L., and Norman E. Bowie. 1983. *Ethical Theory and Business*. 2d ed. Englewood Cliffs, N.J.: Prentice-Hall.

Beck, John C., under the supervision of Bruce R. Scott. 1986. *Japan as a 'Rich Country'* (case). Boston: Harvard Business School.

Benedict, Ruth. 1946. *The Chrysanthemum and the Sword: Patterns of Japanese Culture*. Boston: Houghton Mifflin.

Bennett, John W., and Solomon B. Levine. 1976. "Industrialization and Social Deprivation: Welfare, Environment, and the Postindustrial Society in Japan." In *Japanese Industrialization and Its Social Consequences*, edited by Hugh Patrick. 439–92. Berkeley: University of California Press.

Berman, Daniel M. 1978. *Death on the Job: Occupational Safety and Health Struggles in the United States*. New York: Monthly Review Press.

Bernstein, Aaron, Dan Cook, Pete Engardio, and Gregory L. Miles. 1986. "The Difference Japanese Management Makes." *Business Week*, July 14, 47–50.

Bohnaker, William. 1990. *The Hollow Doll: A Little Box of Japanese Shocks*. New York: Ballantine.

Borrus, Amy. 1990. "How Real Estate Pumps Up Japanese Stocks." *Business Week*, Feb. 12, 78–79.

Bowie, Norman E. 1991. "Challenging the Egoistic Paradigm." *Business Ethics Quarterly* 1(1): 1–21.

Brauchli, Marcus W. 1988. "With All That Extra Free Time, Maybe They'll Seek Second Jobs." *Wall Street Journal*, Dec. 27, B1.

Brauchli, Marcus W., and Clay Chandler. 1990. "Battered Nikkei: Japan's Stock Market, after Defying Gravity, Comes Back to Earth." *Wall Street Journal*, Aug. 23, A–1, A–4.

Brickman, Ronald, Sheila Jasanoff, and Thomas Ilgen. 1985. *Controlling Chemicals: The Politics of Regulation in Europe and the United States*. Ithaca, N.Y.: Cornell University Press.

Brown, Clair, and Michael Reich. 1989. "When Does Union-Management Cooperation Work? A Look at NUMMI and GM–Van Nuys." *California Management Review* 31 (Summer): 26–44.

Burgess, John. 1987. "Japan Says Sayonara to State Railroad." *Washington Post*, April 1, F3.

Burrough, Bruce, and Seth H. Lubove. 1986. "Credibility Gap: Some Concerns Fudge Their Safety Records to Cut Insurance Costs." *Wall Street Journal*, Dec. 2, 1, 28.

Business Roundtable. 1981. "Statement on Corporate Social Responsibility," New York.

Business Week. 1980. "The Reindustrialization of America." June 30, 55–138.

———. 1988. "Japan's Influence in America." July 11, 64–75.

Butler, Richard J., and John D. Worrall. 1983. "Workers' Compensation Benefits and Injury Claims Rates in the Seventies." *Review of Economics and Statistics* 65 (Nov.): 580–89.

Choate, Pat. 1990. *Agents of Influence*. New York: Alfred A. Knopf.

Cole, Robert E., and Donald R. Deskins, Jr. 1988. "Racial Factors in the Employment Patterns of Japanese Auto Firms in America." *California Management Review* 31 (Fall): 9–22.

Committee for Economic Development. 1971. *Social Responsibilities of Business Corporations*. New York.

Deutsch, Morton. 1973. *The Resolution of Conflict: Constructive and Destructive Processes*. New Haven, Conn.: Yale University Press.

Donaldson, Thomas. 1982. *Corporations and Morality*. Englewood Cliffs, N.J.: Prentice-Hall.

Drucker, Peter. 1946. *The Concept of the Corporation*. New York: John Day.

———. 1981a. "Behind Japan's Success." *Harvard Business Review* 59 (Jan.–Feb.): 83–90.

———. 1981b. "What Is 'Business Ethics'?" *Public Interest* 63 (Spring): 18–36.

Emery, F. E., and E. L. Trist. 1965. "The Causal Texture of Organizational Environment." *Human Relations* 18: 21–32.

Foust, Dean, with Wendy Yellner, James B. Treece, and Amy Borrus. 1989. "The UAW vs. Japan: It's Showdown Time in Tennessee." *Business Week*, July 24, 64–65.

Frederick, William C. 1986. "Toward CSR3: Why Ethical Analysis Is Indispensable and Unavoidable in Corporate Affairs." *California Management Review* 28 (Winter): 126–41.

Freeman, R. Edward, and Daniel R. Gilbert. 1988. *Corporate Strategy and the Search for Ethics*. Englewood Cliffs, N.J.: Prentice-Hall.

Friedman, Milton. 1970. "The Social Responsibility of Business Is to Increase Its Profits." *New York Times Magazine*, Sept. 13, 33, 122–26.

Fucini, Joseph L., and Suzy Fucini. 1990. *Working for the Japanese: Inside Mazda's American Auto Plant*. New York: Free Press.

Fujimoto, Takeshi. 1967. "A Short History of Occupational Accidents in Japanese Industries." *Report of the Institute for the Science of Labour* 67: 1–19.

———. 1986. Interview with author, Kawasaki, Japan, Nov. 18.

Fujita, Yoshitaka. 1984. *Employee Benefits and Industrial Relations*. Japanese Industrial Relations Series 12. Tokyo: Japan Institute of Labour.

Gersuny, Carl. 1981. *Work Hazards and Industrial Conflict*. Hanover, N.H.: University Press of New England.

Halberstam, David. 1986. *The Reckoning*. New York: William Morrow.

Hampton, William J. 1988. "How Does Japan Inc. Pick Its American Workers?" *Business Week*, Oct. 3, 84, 88.

Hazama, Hiroshi. 1976. "Historical Changes in the Life Style of Industrial Workers." In *Japanese Industrialization and Its Social Consequences*, edited by Hugh Patrick, 21–51. Berkeley: University of California Press.

Heinrich, Herbert W. 1941. *Industrial Accident Prevention: A Scientific Approach*. New York: McGraw-Hill.

Helm, Leslie, with Charles Gaffney. 1986. "The High Price Japanese Pay for Success." *Business Week*, April 7, 52–54.

Holmes, Sandra. 1976. "Executive Perceptions of Corporate Social Responsibility." *Business Horizons* 19 (June): 34–40.

Holstein, William J., with Pete Engardio and Dan Cook. 1986. "Japan, U.S.A." *Business Week*, July 14, 44–55.

Holusha, John. 1989. "No Utopia, but to Workers It's a Job." *New York Times*, Jan. 29, 1, 10.

Honda Motor Co., n.d. *There Can Be No Production without Safety*. Tokyo.

Iino, Hirohita, and Machida Seiji. 1986. Interviews, Tokyo, Nov. 5, 14.

Imai, Masaaki. 1981. "Has Japan Gone Beyond Capitalism?" *When in Japan: An Introduction for Doing Business in and with Japan* 3: 3–5.

Ingrassia, Paul. 1990. "Toyota to Build Plant in U.S., Reflecting Gains." *Wall Street Journal*, Sept. 6, A-3, A-6.

International Labour Office. 1983. *1982 Year Book of Labour Statistics*. Geneva, Switzerland.

———. 1986. *1985 Year Book of Labour Statistics*. Geneva, Switzerland.

———. 1988. *1987 Year Book of Labour Statistics*. Geneva, Switzerland.

———. 1989. *1988 Year Book of Labour Statistics*. Geneva, Switzerland.

———. 1990. *1989–90 Year Book of Labour Statistics*. Geneva, Switzerland.

Japan Industrial Safety and Health Association. 1984. *1984 Annual Report*. Tokyo.

———. 1985. *General Guidebook on Industrial Safety*. Compiled by Labor Standards Bureau, Ministry of Labor. Tokyo.
———. n.d. *1985/1986 Annual Report*. Tokyo.
———. 1986. *General Guidebook on Industrial Safety*. Compiled by Labor Standards Bureau, Ministry of Labor. Tokyo.
———. 1988a. *1988 Annual Report*. Tokyo.
———. 1988b. *Safety and Health Data Book in Japan*. Tokyo.
———. n.d. *1988/1989 Annual Report*. Tokyo.
———. 1989. *Safety and Health Data Book in Japan, 1989 edition*. Tokyo.
———. 1990. *Safety and Health Data Book in Japan*. Tokyo.
Japan Information Center. 1989. "Antismoking Campaign Gaining Ground in Japan." *Japan Report* 35 (Aug.): 5.
Japan Institute of Labour. 1982. *Industrial Safety and Health*. Japanese Industrial Relations Series 9. Tokyo.
———. 1983. *Labor Unions and Labor-Management Relations*. Japanese Industrial Relations Series 2. Tokyo.
———. 1988. *Industrial Safety and Health*. Japanese Industrial Relations Series 9. Tokyo.
———. 1989. "Working Conditions and the Labor Market." *Japan Labor Bulletin* 28 (Feb.): 2–3.
Japan International Cooperation Agency. 1985. "Outline of Workmen's Accident Insurance Scheme in Japan." Tokyo: Ministry of Labor. Typescript.
Japan Times. 1988. "Labor Minister Calls for Shorter Working Hours." March 31, 2.
Jidosha, Safety and Work Environment Division. 1986. "The Current Management on Safety, Hygiene, Health and Security." Company document. Feb.
Jidosha Workers' Union. 1986. *Union Activities at Jidosha*.
Kamata, Satoshi. 1982. *Japan in the Passing Lane*. Translated and edited by Tatsuru Akimoto. New York: Pantheon.
———. 1986. Interview with author, Tokyo, Dec. 8.
Kaneko, Tsuguo. 1986. Interview. Tokyo. Nov. 17.
Kelman, Steven. 1981. *Regulating America, Regulating Sweden: A Comparative Study of Occupational Safety and Health Policy*. Cambridge, Mass.: MIT Press.
Kendall, Richard M. 1987. "Safety Management: Japanese-Style." *Occupational Hazards* 49 (Feb. 8): 48–51.
Kraar, Louis. 1989. "Japan's Gung-Ho U.S. Car Plants." *Fortune*, Jan. 30, 98–100, 104, 106, 108.
Lee, Gloria L., and John Wrench. 1980. "Accident-Prone Immigrants: An Assumption Challenged." *Sociology* 14 (Nov.): 551–56.
Lee, Sang M., and Gary Schwendiman. 1982. *Japanese Management: Cultural and Environmental Considerations*. New York: Praeger.
Lehner, Urban C. 1988. "Changed History: More Japanese Deny Nation Was Aggressor during World War II." *Wall Street Journal*, Sept. 8, 1, 24.
Lehner, Urban C., and Alan Murray. 1990. "Strained Alliance: 'Selling of

America' to Japanese Touches Some Very Raw Nerves." *Wall Street Journal*, June 19, A–1, A–4.

Levine, Solomon B. 1981. "Labor Regulations." In *Business and Society in Japan: Fundamentals for Businessmen*, edited by Bradley M. Richardson and Taizo Ueda, 52–61. New York: Praeger.

Levitt, Theodore. 1958. "The Dangers of Corporate Social Responsibility." *Harvard Business Review* 36 (Sept.-Oct.): 41–51.

Littleboy, David J. 1989. "Life in Asia Is No Bowl of Cherry Blossoms." *Business Week*, Feb. 6, 7, 9.

Lofgren, Don J. 1989. *Dangerous Premises: An Insider's View of OSHA Enforcement*. Ithaca, N.Y.: ILR Press.

McCaffrey, David P. 1982. *OSHA and the Politics of Health Regulation*. New York: Plenum Press.

McCaffrey, David P., David F. Andersen, Paul McCold, and Doa Hoon Kim. 1985. "Modeling Complexity: Using Dynamic Simulation to Link Regression and Case Studies." *Journal of Policy Analysis and Management* (2): 196–216.

McKean, Margaret A. 1981. *Environmental Protest and Citizen Politics in Japan*. Berkeley: University of California Press.

McMillan, Charles J. 1985. *The Japanese Industrial System*. Berlin, Germany: Walter de Gruyter.

McVoy, Edgar C. 1955. "Introducing Democratic Employment Practice in Japan." In *Hands across Frontiers: Case Studies in Technical Cooperation*, edited by Howard M. Teaf, Jr., and Peter G. Franck, 412–67. Ithaca, N.Y.: Cornell University Press.

———. 1985. "Forced Social Change: Japan's Labor Policies." *International Review of Sociology* 21 (April-Dec.): 25–46.

———. 1986. "A Personalized Study of Social Change: Labor Organization in Japan, 1946–1986." Typescript.

Mafune, Yonosuke. 1986. Interview, Tokyo, Nov. 19.

———. 1988. "Corporate Social Performance and Policy in Japan." In *Research in Corporate Social Performance and Policy*, edited by Lee E. Preston, 291–303. Greenwich, Conn.: JAI Press.

Matsuda, Yasuhiko. 1986. Interview, Tokyo, Nov. 13.

Maynard, Micheline. 1990. "Overtaking the Big Three on Their Own Turf." *USA Today*, Sept. 11, 1B–2B.

Mendeloff, John. 1979. *Regulating Safety: An Economic and Political Analysis of Occupational Safety and Health Policy*. Cambridge, Mass.: MIT Press.

Ministry of Labor. Labor Standards Bureau. 1973. *Dictionary of Labor Terms*. Tokyo.

———. 1980. *Labor Laws of Japan*. Tokyo: Institute of Labor Policy.

———. Industrial Safety and Health Department. Industrial Health Division. 1986. *Working Environment Measurement System in Japan*. Tokyo: Japan Association for Working Environment Measurement.

Mintz, Benjamin W. 1984. *OSHA: History, Law and Policy*. Washington, D.C.: Bureau of National Affairs.

Mirer, Franklin E. 1989. "Prevention of Repeated Trauma Disorders in the Workplace." Testimony before U.S. House of Representatives Committee on Government Operations, Employment and Housing Subcommittee, June 6.

Monden, Yasuhiro. 1983. *Toyota Production System: Practical Approach to Production Management*. Norcross, Ga.: Industrial Engineering and Management Press, Institute of Industrial Engineers.

Nader, Ralph. 1965. *Unsafe at Any Speed*. New York: Grossman.

National Defense Council for Victims of Karōshi. 1990. *Karōshi: When the "Corporate Warrior" Dies*. Tokyo: Mado-Sha.

National Research Council. 1982. *The Competitive Status of the U.S. Auto Industry: A Study of the Influences of Technology in Determining International Industrial Competitive Advantage*. Washington, D.C.: National Academy Press.

National Safety Council. 1983. *Work Injury and Illness Rates 1983*. Chicago.

———. 1990. *Work Injury and Illness Rates 1989*. Chicago.

Nelkin, Dorothy, and Michael S. Brown. 1984. *Workers at Risk*. Chicago: University of Chicago Press.

Noble, Charles. 1986. *Liberalism at Work: The Rise and Fall of OSHA*. Philadelphia: Temple University Press.

Ouchi, William G. 1981. *Theory Z: How Americans Can Meet the Japanese Challenge*. Reading, Mass.: Addison-Wesley.

———. 1984. "Political and Economic Teamwork: The Development of the Microelectronics Industry in Japan." *California Management Review* 26 (Summer): 8–34.

Page, Joseph A., and Mary-Win O'Brien. 1973. *Bitter Wages: Ralph Nader's Study Group Report on Disease and Injury on the Job*. New York: Grossman.

Parker, Mike. 1990. Interview, Detroit, Nov. 12.

Parker, Mike, and Jane Slaughter. 1988. *Choosing Sides: Unions and the Team Concept*. Boston: South End Press.

Pascale, Richard T., and Anthony G. Athos. 1981. *The Art of Japanese Management: Applications for American Executives*. New York: Warner Books.

Patrick, Hugh, ed. 1976. *Japanese Industrialization and Its Social Consequences*. Los Angeles: University of California Press.

Pempel, T. J. 1990. Personal Correspondence, Oct. 30.

Pharr, Susan J., and Joseph L. Badaracco. 1986. "Coping with Crisis: Environmental Regulation." In *America versus Japan*, edited by Thomas G. McGraw, 229–59. Boston: Harvard Business School Press.

Prestowitz, Clyde V., Jr. 1989. *Trading Places: How We Are Giving Our Future to Japan and How to Reclaim It*. New York: Basic Books.

Pummer, Christopher. 1986. "Injured Workers Await a Remedy at Chrysler." *Wilmington Sunday News Journal*, March 16, C1, C6.

Roanoke Times and World News. 1988. "Inspections Bogus, OSHA Workers Say." April 20, B5.

———. 1989. "Buying Land in Tokyo, a Pricey Proposition." April 16, A11.

Rosner, David, and Gerald Markowitz, eds. 1987. *Dying for Work: Workers'*

Safety and Health in Twentieth Century America. Bloomington: Indiana University Press.

Ruble, Thomas L., and Kenneth W. Thomas. 1976. "Support for a Two-Dimensional Model of Conflict Behavior." *Organizational Behavior and Human Performance* 16 (1): 143–55.

Samuels, Howard. 1986. Interview, Washington, D.C., May 16.

Schonberger, Richard J. 1982. *Japanese Manufacturing Techniques*. New York: Free Press.

Sethi, S. Prakash. 1975. *Japanese Business and Social Conflict*. Cambridge, Mass.: Ballinger.

Sethi, S. Prakash, Nobuaki Namiki, and Carl L. Swanson. 1984. *The False Promise of Japanese Management*. Boston: Pittman.

Shapiro, Margaret. 1988. "Old Black Stereotypes Find New Lives in Japan." *Washington Post*, July 22, A18.

Sherman, Stratford P. 1991. "Japan's Influence on American Life." *Fortune*, June 17, 115–24.

Shimada, Haruo. 1980. *The Japanese Employment System*. Japanese Industrial Relations Series 6. Tokyo: Japan Institute of Labour.

Simon, Philip J. 1983. *Reagan in the Workplace: Unraveling the Health and Safety Net*, edited by Kathleen Hughes. Washington, D.C.: Center for the Study of Responsive Law.

Smith, Lee. 1990. "Fear and Loathing of Japan." *Fortune*, Feb. 26, 50–65.

Sōhyō. 1985. *This Is Sōhyō: Japanese Workers and Their Struggles*. Tokyo.

Solo, Sally. 1990. "Japan's Unhappy Auto Workers." *Fortune*, Oct. 22, 10–11.

Steiner, George A., and John F. Steiner. 1985. *Business, Government, and Society: A Managerial Perspective*. 4th ed. New York: Random House.

Tanabe, Hajime. 1984. *Practical Manual on Hazards Prediction Activities*. Tokyo: Japan Industrial Safety and Health Association.

Tamura, Toshiyuki, and Sohtaro Kunihisa. 1982. "Education, Transport Capital and Productivity Change: The Case for Japan's Experience since the Meiji Era." In *Japanese Management: Cultural and Environmental Considerations*, edited by Sang M. Lee and Gary Schwendiman, 234–47. New York: Praeger.

Taylor, Jared. 1983. *Shadows of the Rising Sun: A Critical View of the "Japanese Miracle."* New York: William Morrow.

Thomas, Kenneth W. 1976. "Conflict and Conflict Management," In *Organizational Psychology*, edited by Marvin Dunnette, 889–935. Chicago: Rand McNally.

———. 1979. "Organizational Conflict," In *Organizational Behavior*, edited by Steven Kerr, 899–935. Columbus, Ohio: Grid.

Tiffany, Paul. 1987. "Japanese Corporate Lobbying in the United States: Strategy and Structure." Working Paper 87–11. Philadelphia: Reginald H. Jones Center for Management Policy, Strategy and Organization, Wharton School of Business, University of Pennsylvania.

Tomihisa, Osamu, and Yoko Yoshida. 1986. Interview, Tokyo, Oct. 28.

Treece, James B. 1988. "Will the Auto Glut Choke Detroit?" *Business Week*, March 7, 54–62.

Tsurumi, Yoshi. 1978. *Japanese Business: A Research Guide with Annotated Bibliography*. New York: Praeger.
Ueshima, Koki. 1989. "Firms Extend Care beyond Workers' Lifetimes." *Japan Economic Journal*, April 1, 8.
United Auto Workers Health and Safety Department. 1989. "Analysis: Occupational Injury and Illness Ratios, Motor Vehicle Industry, 1983–1987." Unpublished report. Sept.
United Nations. 1988. *1986 Demographic Yearbook*. New York.
U.S. Congress. Office of Technology Assessment. 1985. *Preventing Illness and Injury in the Workplace*. Washington, D.C. April.
U.S. Department of Labor. Bureau of International Labor Affairs. 1989. *Foreign Labor Trends: Japan*. Prepared by American Embassy, Tokyo. Washington, D.C.: U.S. Government Printing Office.
U.S. Department of Labor. Bureau of Labor Statistics. 1977. *Handbook of Labor Statistics, 1977*. Washington, D.C.: U.S. Government Printing Office.
———. 1982. *Occupational Injuries and Illnesses in the United States by Industry, 1980*. Washington, D.C.: U.S. Government Printing Office.
———. 1987. *Occupational Injuries and Illnesses in the United States by Industry, 1985*. Washington, D.C.: U.S. Government Printing Office.
———. 1988. *Occupational Injuries and Illnesses in the United States by Industry, 1986*. Washington, D.C.: U.S. Government Printing Office.
———. 1989. *Occupational Injuries and Illnesses in the United States by Industry, 1987*. Washington, D.C.: U.S. Government Printing Office.
———. 1990a. *International Comparisons of Hourly Compensation Costs for Production Workers in Manufacturing, 1975–1989*. Report 794. Washington, D.C.
———. 1990b. *Occupational Injuries and Illnesses in the United States by Industry, 1988*. Washington, D.C.: U.S. Government Printing Office.
U.S. President. 1981. *Report of the President to the Congress on Occupational Safety and Health for Calendar Year 1980*. Washington, D.C.: U.S. Department of Labor, Occupational Safety and Health Administration.
———. 1984. *Report of the President to the Congress on Occupational Safety and Health for Calendar Year 1983*. Washington, D.C.: U.S. Department of Labor, Occupational Safety and Health Administration.
———. 1987. *Report of the President to the Congress on Occupational Safety and Health for Calendar Year 1986*. Washington, D.C.: U.S. Department of Labor, Occupational Safety and Health Administration.
———. 1988. *Report of the President to the Congress on Occupational Safety and Health for Calendar Year 1987*. Washington, D.C.: U.S. Department of Labor, Occupational Safety and Health Administration.
Viscusi, W. Kip. 1983. *Risk by Choice*. Cambridge, Mass.: Harvard University Press.
Vogel, David. 1986. *National Styles of Regulation: Environmental Policy in Great Britain and the United States*. Ithaca, N.Y.: Cornell University Press.
Vogel, Ezra F. 1985. *Comeback*. New York: Simon and Schuster.
Wallick, Franklin. 1972. *The American Worker: An Endangered Species*. New York: Ballantine.

Wall Street Journal. 1990. "Japanese Auto Workers Seek Longer New-Model Cycles." Sept. 18, A–22.

Webb, Eugene J., Donald T. Campbell, Richard D. Schwartz, and Lee Sechrest. 1966. *Unobtrusive Measures: Nonreactive Research in the Social Sciences*. Chicago: Rand McNally.

Weidenbaum, Murray L. 1981. *Business, Government and the Public*. 2d ed. Englewood Cliffs, N.J.: Prentice-Hall.

Whiting, Robert. 1977. *The Chrysanthemum and the Bat: Baseball Samurai Style*. New York: Dodd, Mead.

Wilson, Graham K. 1990. *Business and Politics: A Comparative Introduction*. Chatham, N.J.: Chatham House.

Wokutch, Richard E. 1990a. *Cooperation and Conflict in Occupational Safety and Health: A Multination Study of the Automotive Industry*. New York: Praeger.

———. 1990b. "Corporate Social Responsibility Japanese Style." *Academy of Management Executive* 4 (May): 56–74.

Wokutch, Richard E., and Josetta McLaughlin. 1988. "The Sociopolitical Context of Occupational Injuries." In *Research in Corporate Social Performance and Policy*, edited by Lee E. Preston, 10: 113–37. Greenwich, Conn.: JAI Press.

Womack, James P., Daniel T. Jones, and Daniel Roos. 1990. *The Machine That Changed the World*. New York: Rawson Associates.

Yamaguchi, Yuzo. 1991. "Dying for a Job." *Economic World* 23 (April): 34–36.

Yates, Brock. 1983. *The Decline and Fall of the American Automobile Industry*. New York: Empire Books.

Yoshimoto, Minoru. 1981. *Details of the Labor Safety and Health Law*. Tokyo: Association of Labor Law and Legislation.

Index

Absenteeism, 98, 99, 152, 241*n*1
Abused Japanese worker, myth of, 4, 9–10, 202, 224–25, 226
Accident investigations, 115–16, 173
Activities, safety and health at Jidosha Japan, 109–20
 prestart-up, 163–70
Administrative guidance, Japan's emphasis on, 8, 64, 72
Affiliation, group, 32, 51–52. *See also* Loyalty, Japanese sense of
Aging of population, 38, 74–75, 118
Agricultural support policies, 37
Alcohol abuse, 117–18, 240*n*11
 drunk driving laws, 117
All Nippon Airways, 52
All-terrain vehicle manufacturers, 50, 53
Altruism
 corporate social responsibility and, 15, 17, 232
 economic or financial vs. traditional notion of, 237*n*5
Amakudari, 39
Anti-smoking activities, 116–17
Asbestos in workplace, use of, 75
Assertiveness
 conflict-coping behavior and, 19, 79–80, 81
 contingency theory of cooperation and, 228, 229
 in U.S., 144, 231
Attendance bonus system in Jidosha USA, 151, 152
Authority, American vs. Japanese views of, 169
Automotive industry, 11–13, 79
 Big Three auto firms, 12, 106, 171
 fatality rates for U.S. vs. Japan, 209–10, 212–13
 injury and illness rates in
  in Japan, 121, 123, 124, 125, 219–20, 221
  in U.S., 192–94, 195
 See also Jidosha, occupational safety and health at; Jidosha USA, occupational safety and health at
Avoidant behavior, 18, 19
Awards, safety and health, 116, 119

Bacow, Lawrence S., 7
Badaracco, Joseph L., Jr., 8
Benefits, employee, 39–40, 97, 101
Bhopal, Union Carbide tragedy in, 7, 231
Bias in comparative injury and illness data, sources of, 196–204
Bieber, Owen, 12–13
Black lung disease, 61

Candor, Japanese concept of, 11
Carpal tunnel syndrome, 186
Charity, sense of, 52
Chrysler Corporation, 11, 12
Citations, percentage of inspections resulting in, 90–91
Civil servants, prestige of career, 45–46, 93
Collaborative behavior, 18, 19
Committees, safety and health, 71, 110–11, 170–71, 178
Compensation
 basis of, 40
 executive, U.S. vs. Japanese, 225
 at Jidosha USA, 151
 See also Wages; Workers' compensation
Competition, 18, 19
 among auto industry firms in Japan, 220

cooperativeness of labor in Japan and, 42–43
Compliance
Japanese penchant for, 53
work-rule, 77, 139, 189, 229–30
Conflict, 17–19
defined, 17
in Japanese occupational safety and health system, 76–81
at Jidosha Japan, 134–44
at Jidosha USA, 177–84
two-dimensional model of conflict behavior, 18–19
Conformity, 53, 66, 144, 189
Confucian ethics, 52
Conservation (Japan), 37–38
Constitution, Japanese, 57
Consumer movement in Japan, 49
Cooperation, 17–19
among auto industry firms in Japan, 220
contingency theory of, 228–30
defined, 17
Industrial Safety and Health Law reliance on, 66
in Japanese occupational safety and health system, 76–81
at Jidosha Japan, 134–44
at Jidosha USA, 173–84
labor-management, 41, 42–43
Corporate culture, transmission from Jidosha Japan to Jidosha USA, 169
Corporate social responsibility (CSR), 5, 13–17
competitive implications of, 232–35
divergence between reality and image for Japanese firms, 234
motivations and determinants of, 47–54
Corporation as paternalistic organization, 39–40
Cross-national adaptation, potential for, 223, 224
Cross-ownership of stock, 49
Cultural differences
comparative performance and, 188
injury and illness rate discrepancies and, 194
Cumulative trauma disorders (CTDs), 73, 186, 228, 229, 242*n*7
in Japanese transplants, 223
at Jidosha USA, 161, 162, 179–80, 189–90
compared to Jidosha Japan, 191–92
compared to U.S. auto industry, 193–94, 195
underreporting of, 77

Decision making, 67, 109, 162–63, 168
Democratic Socialist party, 44
Demographic conditions in Japan, 36–37, 38
Deutsch, Morton, 18
Diet, the (Japanese legislature), 45, 60
Discharge of personnel, 99–100
Discipline for violations, 139–40, 151
Disease control, 116
Division-level safety and health committee, 111
Down time, activities required during, 143, 167
Drucker, Peter, 11
Drug abuse, 117–18
Drug and alcohol testing, 117, 157
Dust in workplace, legislation on, 67

Economic conditions in Japan, 36–38
predictions of growth in 1990s, 46
Employee assistance program at Jidosha USA, 157–58
Employer associations, 44, 67
Employment Security Law of 1947, 36
Endaka (high-yen condition), 46
Enforcement activities in U.S. vs. Japan
inspections, 65–66, 83–88
staffing, 81–83, 85, 87, 93, 187–88
violations and penalties, 64, 88–94
Enlightened self-interest (ESI), 15–16, 50–51, 53–54, 227, 232
Enterprise unionism, 39
Environment, exposure levels for harmful substances in work, 67–68
Environmental conditions in Japan, 25–54
demographic, geographic, and economic conditions, 36–38
historical factors, 30–36
labor-management-government relations, 38–45
manufacturing management system, 25–30
organizational conditions, 25–30
political and governmental conditions, 45–47
Environmental disasters, 48
Environmental movement, 47–48
Equal Employment Opportunity Law (1986), 42, 49
Ergonomics at Jidosha USA, 179
Establishment size, work injuries and illnesses and, 201, 215–16
Ethics, social norms and, 51–54
Ethnic composition, 153, 154, 204
Executive compensation, U.S.-Japanese cost disparities and, 225
Executive Office for Workmen's Accident Compensation Insurance, 59
Exercise, 116, 155
External political pressures, 46–47

Factory Law of 1911, 33, 56
Factory systems, evolution of, 33
Fatalities, worker
standard company procedure after, 103
U.S.-Japan rate comparisons, 191, 198, 201–2, 209–15
Federation of Japanese Automobile Workers' Unions, 44, 74, 143
Feudal era, values and customs from, 31–32, 40
Bushido, 52
Fines for safety and health violations, 66–67, 125
Five-Year Industrial Injury Prevention Program, 58–59
Flexible manufacturing systems, Japanese, 26, 30
Ford, Henry, 26
Ford Motor Company, 12
Fraser, Douglas, 12–13
Friedman, Milton, 14, 15
Fujimoto, Takeshi, 34

Gaijin, 10
Gary, Elbert, 108
General Guidebook on Industrial Safety (JISHA), 72
General Motors, 12
"Get a Hundred Campaign," 116
Government
conditions in Japan, 45–47
conflict-coping behavior of management toward, 79–80
organization of safety and health activities, 59–60
prestige of career in, 39, 45–46, 93
promotion of safety and health beyond regulations, 68.
See also Labor-management-government relations
Grievance procedures, 142, 151–52
Group decision making, 168
Group identification, sense of, 32, 51–52
Groups, American difficulties working in, 168

Handicapped workers, 98
Harmful substances in work environment, exposure levels for, 67–68
Hazard-prediction activities (KYT activities), 112–13, 160, 172
Health insurance at Jidosha Japan, 98
Health Insurance Law of 1922, 56
Health of workers at Jidosha, statistics on, 126–27
Heinrich, Herbert W., 213
Historical conditions in Japan, 30–36
Hitachi, blunders in U.S. operations of, 50
Holidays, 98
Holiday sickness, 74
Homogeneity of Japanese population, 204
physical similarities, compared to U.S. workers, 188
Honda Motor Company, 28–30, 39, 46, 50
Four Cs, 28, 30
Honne, 11, 234
Hospitals, company, 39–40, 105, 124–25
Housing, company-owned, 39

Iacocca, Lee, 11
Ieyasu, Tokugawa, 32
Illness. *See* Injury and illness, work
Image and reality, divergence between, 234
Immigration, 51, 204
Individualism, 51, 148–49
Individual rights and freedoms, priority in Japan, 66
Industrial composition, reported injury and illness rates and, 203–4
Industrial disputes, workdays lost due to, 43
Industrial Health/Hygiene Academic Association, 67
Industrial Homework Law of 1970, 61
Industrial Injury Prevention Organization Law of 1964, 61, 72
Industrialization, history of Japanese, 31, 32–36
Industrial organization for safety and health, 70–72
Industrial Safety and Health Department, 59
Industrial Safety and Health Law of 1972, 58, 60, 61–66, 136
Injury and illness, work
in construction firms, 75, 204
definitions of, 194
distortions of rates in Japanese auto industry, 124–26
evaluation of managers based on, 182–83
frequency rates, industrial, 207–8
at Jidosha Japan, 120–27
reports and investigations, 113–16
for subcontractors, 127–33
at Jidosha USA, 161–62, 173
U.S. auto industry compared to, 192–93, 195
severity rate
for Jidosha Japan and Japanese auto industry, 121, 123, 124, 125
for Jidosha USA, 190, 191
for subcontractors, 127, 128–32

for ten Japanese automotive firms, 219, 221
U.S. vs. Japan, 209, 210–12
in smaller firms, 75
in U.S. and Japan, 6, 196–223
bases for rate calculations, 201–2
company- and plant-level comparisons, 216–22
data on, 196–204
definitions, 197–98
establishment size and, 201, 215–16
ethnic homogeneity and, 204
fatality rate comparisons, 209–15
industrial composition and, 203–4
industrial relations and, 202–3
national traditions and, 203
political considerations in, 202
rates of, 189–95
reporting incentives and disincentives, 141–42, 198–200
trends, 205–9
U.S. subsidiaries of Japanese automotive firms, 217–19
worker concealment of, 104
See also Underreporting (underrecording) of injuries and illnesses
Inspections
at Jidosha Japan
by Labor Standards Bureau, 137
by union, 120
by workers, 111–12
at Jidosha USA, 171–72
by SOSHA, 174–75
notification in advance of, 66
OSHA, 162
regional differences in priorities for, 68
in U.S. vs. Japan, 65–66, 83–88
worker requests for, 68
Inspectors, Japanese vs. U.S., 65–66, 82–84, 85, 87, 137
Insularity of Japanese businesses, 49
Insurance premiums and reporting disincentives, 198–99
International Labour Organization, 57
Inventory systems, 26, 42
Itai Itai disease, 48

Jackson, Jesse, 154
Japan
external pressures imposed on, 46–47
manufacturing management system in, 25–30
myths about occupational safety and health in, 4–5, 9–10, 202, 224–25, 226
See also Environmental conditions in Japan
Japan Air Lines, 118
Japan Bioassay Laboratory, 68
Japanese Automobile Manufacturers' Association (JAMA), 45, 67, 121
Committee of Safety and Health Managers of, 219–20
Japan Industrial Safety and Health Association (JISHA), 61, 68, 72–73, 75, 95, 107, 112, 113
Japan Institute of Labour, 74, 76, 78, 207
Japan National Railways (JNR), privatization of, 45, 100
Japan Socialist party, 44
Jidōka, notion of, 26, 102, 149
Jidosha, occupational safety and health at
cooperation and conflict in, 134–44
cumulative trauma disorders at, 191–92
"emphasis injuries," 161–62
goals and activities, 106–20
injury and illness experience, 120–27
reports and investigations, 113–16
of subcontractors, 127–33
labor practices, 96–102
organization of safety and health function, 102–6
staffing levels, 105, 106
Jidosha USA, occupational safety and health at
comparative performance, 188–95
cooperation and conflict in, 177–84
hiring practices, 150–51, 154
labor practices, 147–55
preemployment drug and alcohol testing, 117, 157
production system at, 146–47
productivity at, 145
safety and health system, 155–77
current operations, 170–73
goals and objectives, 160–63
organizational structure and staffing, 157–60, 187–88
prestart-up activities, 163–70
relations with regulators, 173–76
union activities, 176–77
U.S. vs. Japanese standards and practices, 184–87
Job-hopping, increase in, 40
Job-rotation system (multiskilled worker program), 101–2, 180–81
Joint consultation scheme, 71
Jones, Daniel T., 11
Just-in-time inventory system, impact on subcontractors, 42

Kaizen activities, 27, 101, 109, 143, 168
Kamata, Satoshi, 143
Kanban inventory system, 26
Karōshi (death from overwork), 4, 70, 74
Kelman, Steven, 21

Kunihisa, Sohtaro, 32
KYT (*Kiken Yochi* Training) activities, 112–13, 160, 172

Labor hours per car, U.S. vs. Japanese, 12
Labor Insurance Appeals Committee, 69
Labor-management-government relations
contingency theory of cooperation in, 228–30
in Japan, 38–45
at Jidosha, cooperative and conflictual aspects of, 134–44
in U.S. compared to, 184–85
in U.S. auto industry, 12–13
Labor-management relations
adversarial, in U.S., 182, 202–3
cooperation, 41, 42–43
reported injury and illness rates and, 202–3
Labor practices
at Jidosha Japan, 96–102
at Jidosha USA, 147–55
Labor Relations Adjustment Act of 1946, 36
Labor Standards Bureau (Japan), 28, 62, 64, 73, 95, 144, 228, 232
broad focus, 82
enforcement activities, 90–94
inspection activities, 85, 86, 88
inspectors, 65–66, 137
legislation initiated by, 67
organization of, 59–60
prestige of, 93
special committees, 136
staffing, 82–84, 85
workers' compensation administered in, 68–70
Labor Standards Law of 1947, 36, 57–58
Land, scarcity and value of Japanese, 36–38, 138
Lean production, 26
Legislation. *See* Occupational safety and health system, Japanese
Liberal Democratic party (LDP), 37, 45, 144
Lifestyles, promotion of healthful, 116–17
Lifetime employment, 39, 40–41, 100, 134
Line managers, responsibility for safety and health, 70, 102–5, 168
Lockheed payoff scandal, 52
Lost-time incidence rate
of Japanese automotive transplants, 217–22
for Jidosha and Japanese auto industry, 121–23, 124
for Jidosha subcontractors, 127, 128–32
at Jidosha USA, 190, 191
compared to U.S. auto industry, 192, 193
reporting of, 114
differences in criteria used for, 197
for ten Japanese auto firms, 219–20
Loyalty, Japanese sense of, 40, 52, 76, 203

MacArthur, Douglas, 35
McVoy, Edgar, 34
Machine That Changed the World, The, (Womack, Jones, and Roos), 11
Mafune, Yonosuke, 49
Management
bonus plans for, 151, 182–83
union officials' ability to move into, 41, 134–35
Managers
American, 153
evaluation of, on safety and health performance, 104–5
safety and health, prestige of, 75
Manufacturing management system in Japan, 25–30
American system vs., 26, 231
Manufacturing sector, injury and illness rates in U.S. vs. Japan, 210–13
Marshall v. *Barlow Inc.*, 66
Master employers, 39–41
permanent employees of, 40–41, 53, 226
safety and health conditions in, 78, 79
secondary classes of workers within, 42
services provided to employees by, 39–40
subcontractors and, 42, 77–78, 114–15, 136, 227
Medical examinations, 116
Meiji Revolution (1868), 31, 32
Memorandum of Understanding on Safety and Health (1973), 176
Mendeloff, John, 6–7
Mental health care, 118
Methodology, 22–23
Microchip manufacturing, toxic chemicals in, 75
Militancy, union, 36
Minamata disease, outbreak of, 48
Mine Law of 1905, 33, 56
Ministerial ordinances, 60–61
Ministry of Industry (Kobushō), 33
Ministry of International Trade and Industry (MITI), 38, 39
Ministry of Labor, 35–36, 46, 59, 68, 200, 208
Minority employment in Jidosha USA, 153, 154
Minority rights movements, 48
Modernization, policies for, 32–36

Monitoring of workshops for hazards, 116
Morning exercises, 155
Morning safety check, 112–13, 172
Multinationals in developing countries, 232
Myths about occupational safety and health in Japan, 4–5, 9–10, 202, 224–25, 226

Nader, Ralph, 12
Nakasone, Yasuhiro, 45, 50, 153–54
Narita Airport, 48
National Ability Report Law (1943), 34
National Institute of Industrial Health, 68
National Labor Patriotic Cooperation Law (1941), 34
National safety and health week, 57
National Safety Council (U.S.), 6, 72, 207–8
Near-miss accident reporting system, 161, 171–72, 178
Nenko wage system, 39
New Directions movement, 177, 182, 183
Nissan, 20, 39
Nobility (daimyō), 31
Norms, ethics and social, 51–54

Obedience, American vs. Japanese views regarding, 169
Occupational Health Service Center, 68
Occupational safety and health
  altruistic motivations for, 15, 17, 232
  conditions in developing countries, 227, 232
  conflict between Japanese manufacturing system and approach to, 30
  controversial nature of, 10
  enlightened self-interest and, 16
  importance of, 5–9
  international and political implications of, 9–11
  in Japan, myths about, 4–5, 9–10, 202, 224–25, 226
  Japanese vs. American attitudes toward, 93, 136, 181–82, 230
  See also Environmental conditions in Japan; Jidosha, occupational safety and health at; Jidosha USA, occupational safety and health at
Occupational Safety and Health Act, U.S. (1970), 33, 62–66, 205
Occupational Safety and Health Administration (OSHA), 82, 228
  cooperative programs under Reagan administration, 230
  costs of compliance with, 7
  criticism of, 6–8
  enforcement activities, 88–94
  inspection activities, 86–88, 162
    percentage resulting in citations and penalties, 90–91
  lack of respect for, 93, 136
  staffing, 82, 87
  See also State OSHA programs (SOSHA)
Occupational Safety and Health Review Commission (U.S.), 65
Occupational safety and health system, Japanese, 55–94, 226–27
  enforcement activities in U.S. vs. Japan, 81–94
    inspections, 65–66, 83–88
    staffing, 81–83, 85, 87, 93, 187–88
    violations and penalties, 64, 88–94
  regulation, 55–81
    cooperation and conflict in, 76–81
    current problems and concerns, 73–76
    early legislation, 55–57
    industrial organization for safety and health, 70–72
    Industrial Safety and Health Law (1972), 58, 60, 61–66, 136
    organization of government safety and health activities, 59–60
    postwar legislation, 57–59
    recent legislation, 60–61
    rulemaking, enforcement and assistance activities, 66–68
    volunteerism in, 63, 72–73, 93
    workers' compensation, 56, 68–70
Office of Technology Assessment, U.S., 6, 205
Office workers, commitment of, 134
"Oil shocks" of 1973 and 1979, 12
Organizational conditions in Japan, 25–30
Organization of safety and health function, 102–6, 157–60
Overtime work, 74, 97–98, 120, 134, 142
Overwork, death from (*karōshi*), 70, 74

Pace of work, 74, 142–43, 166–67
Paradise, myth of Japanese workplace as, 4–5, 226
Participative decision making, 109, 162
Part-time workers at Jidosha USA, 148, 180, 182
Peer pressure, 189
Penalties in U.S. vs. Japan, 64, 65, 88–94
Performance, safety and health
  comparative, in U.S. and Japanese Jidosha plants, 188–95
  evaluation of managers on, 104–5
  in Japanese transplants, 229–30
Permanent disabilities, 198
Perry, Matthew C., 31, 32
Personal Service Drafting Law (1939), 34

Personal vacation days, 98, 99, 142
Physical stress, 73–74, 223
Planning process at Jidosha USA, 162–63
Plant-level safety and health committees, 110–11
Plant-level safety and health experts, 105–6
Pneumoconiosis Law of 1960, 61
Political conditions in Japan, 45–47
Political significance of injury and illness statistics, 202
Population (Japanese)
 aging of, 38, 74–75, 118
 density, 36, 37
 homogeneity of, 204
Powerlessness, contingency theory of cooperation and, 228
Prayer, safety and health, 108–9, 170
Prefectural labor standards bureaus, 64
Prestart-up activities at Jidosha USA, 163–70
Private sector, injury and illness rates in U.S. vs. Japan, 210–11, 212, 213, 214
Production methods, Japanese
 conflicts between safety and health concerns and, 181–82
 cumulative trauma disorders resulting from, 179
 at Jidosha USA, 146–47
Production workers. *See* Workers
Productivity
 improving through "rationalization," 101
 at Jidosha USA, 145, 176, 178
 relation of safety and health to, 137–38
Product safety issues, 53
Profitability, safety and health and, 137–38
Psychological stress, 73–74, 223
Public opinion of Japanese companies in U.S., 233

Quality
 job rotation and, 180–81
 management of, in U.S. vs. Japanese auto industries, 27
 safety and health requirements and, 138
Quality circles, 101

Racial issues, sensitivity to, 153–54
Reagan administration, 12, 230
Real estate market (Japan), 37
Recruit Cosmos influence-buying scandal, 50
Regulation, Japanese. *See under* Occupational safety and health system, Japanese
Regulators
 Jidosha USA's relations with, 173–76
 See also Labor Standards Bureau; Occupational Safety and Health Administration (OSHA)
Regulatory policy, comparative, 20–22
Reporting of injury and illness
 cases, 200–201
 conflict over, 76–77
 incentives and disincentives in, 198–200
 at Jidosha, 113–16
  rate discrepancies between Jidosha USA and, 194
 near-miss accident reports, 161, 171–72, 178.
 See also Underreporting of injuries and illnesses
Reprimands, written, 139
Research activities, 68, 72–73
Research Institute of Industrial Safety, 68
Resource poverty, 36–38
Restricted work activity cases, 197, 208–9
Ritualistic practices, 108–9
Roos, Daniel, 11
Ruble, Thomas L., 18

Safety and health committees, 71, 110–11, 170–71, 178
Safety and Health Data Book in Japan (JISHA), 72
Safety and Health Department at Jidosha USA, 157, 158
Safety and Health Education Center, JISHA, 72
Safety and health managers, prestige of, 75
Safety check (finger point and call), 112–13, 172
Safety patrols, union, 120
Safety supervisor, 70
Samurai (warrior class), 31, 140
Shame, Japanese sense of, 194, 229
Sharing behavior, 18, 19
Shōgun, 31
Shuntō (Spring Offensive), 43
Signs, use of, 111, 172
SKY (immediate hazard prediction) activities, 113, 172, 240*n*8
Sloganeering, 107, 108–9, 178
Small firms
 injury and illness rates in, 75
 labor-management-government relations in, 80, 81
 worker protection in U.S. vs. Japan, 216, 223, 227
Smoking in Japan, 116–17, 240*n*9
SOAR (Successful Objectives Achieve Rewards), 151
Social contract, new, 19–20
Social movements in Japan, 47–50

Social responsibility. *See* Corporate social responsibility (CSR)
Sony, 39
South Africa, economic relations with, 48
Specialists, lack of professionally trained safety and health, 75
Staffing, safety and health, in U.S. vs. Japan, 81–83, 85, 87, 93
 at Jidosha Japan, 105, 106, 187–88
 at Jidosha USA, 157–60, 187–88
Standards and practices, U.S. vs. Japanese, 184–87
State OSHA programs (SOSHA)
 enforcement activities, 88–94
 inspection activities, 86–88
 Jidosha USA's cooperative relationship with, 173–76
 standards, 184–87
Statistical monitoring of diseases, 116
Stockholders, influence of, 49
Stress, psychological and physical, 73–74, 223
Strikes, 36, 134
Styles of worker protection, U.S. vs. Japanese, 8–9
 behavioral orientation, 9, 65, 71, 72, 109, 110, 169
 engineering orientation, 8–9, 65, 169
 lessons to be learned from, 230–31
Subcontractors
 injury and illness statistics from, 114–15, 127–33
 relationship to master employers, 42, 77–78, 114–15, 136, 227
 safety and health conditions in, 77–79, 110, 227
 sending employees on "loan" to, 100
Subsidiaries, injury and illness statistics from, 114–15
Suicides in Japan, high incidence of, 70, 74, 118
Sumitomo Bank, 50
Suppliers. *See* Subcontractors
Suzuki Motors, 50, 53
Sweden, regulation of occupational safety and health in, 21

Takeshita, Noburu, 50
Tamura, Toshiyuki, 32
Tanaka, Kakuei, 52
Tatemae/honne distinction, 11, 77, 234
Taylor, Jared, 78
Teams, work, 101, 111, 149–50
Temporary workers, 42, 100–101, 138
 part-timers at Jidosha USA, 148, 180, 182
Theory Z management approach, 41
Third World subsidiaries, safety and health conditions in, 227
Thomas, Kenneth W., 18
"Toolbox meetings," 71
TOPS (Team-Oriented Production System), 151
Toshiba, 39, 50
Toyota, 26, 39, 118
Trade Union Act of 1945, 36
Trade unions. *See* Unions
Training, safety and health, 110, 163–70
Transplants, Japanese, 13, 20, 160, 227–28, 237*n*2
 challenges in U.S. for, 50
 injury and illness experience of, 217–19
  cumulative trauma disorders, 223
  vs. traditional U.S. auto firms, 217–22
 safety and health performance in, 229–30
 See also Jidosha USA, occupational safety and health at
Turnover rate, 99–100, 152, 241*n*2
Two-hand control buttons, U.S. vs. Japanese use of, 185

Underreporting (underrecording) of injuries and illnesses, 7, 104, 203, 232
 by American trainees, 169
 confusion about recording requirements and, 205
 in Japan, 76–77, 208
  in auto industry, 125–26
  vs. U.S., 213–15, 228–29
 at Jidosha USA, 182
 pressures involved in, 141–42
Union associations, 43–45
Unions
 assertiveness of U.S. vs. Japanese, 231
 in Japan, 34, 41, 43, 71, 232
  enterprise unions, 39
  at Jidosha Japan, 112, 119–20, 140–42
  legalization of, 36
  officials' ability to move into management, 134–35
 at Jidosha USA, 150–51, 176–77, 183
 New Directions movement, 177, 182, 183
 safety and health representative, 157, 164, 177–78, 183
United Auto Workers (UAW), 79
 at Jidosha USA, 150–51, 152, 167, 176–77, 183
 organizing drive at Nissan (1989), 20
 safety and health representative, 157, 164, 177–78
 and underrecording of injuries and illnesses, 7
United States

challenges for Japanese firms operating in, 50
competitive behavior of management and labor in, 79, 80
engineering orientation of, 8–9, 65, 169
Japanese male reactions to women and minorities in, 51–52
mass production system in, 26
occupation of Japan after WWII, 35, 57–58
punitive approach of, 93
social movements in, 47
work injury and illness experiences. *See* Injury and illness, work
United States auto industry. *See* Automotive industry
U.S. Bureau of Labor Statistics, 200, 205
U.S. National Institute for Occupational Safety and Health (NIOSH), 68
University of Occupational and Environmental Health, 68
Unsafe at Any Speed (Nader), 12

Vacation days, 98, 99, 142
Violations, safety and health
discipline for, 139–40, 151
fines for, 66–67, 125
penalties for, 64, 88–94
Vogel, David, 21, 245*n*1
Voluntary restraint agreements, 13
Volunteerism in Japanese safety and health system, 63, 72–73, 93

Wage Control Ordinance (1939), 34
Wage-replacement payments for disabilities, 199–200
Wages
annual spring negotiations, 43
in Japan vs. U.S., 225
at Jidosha Japan, 96–97
at Jidosha USA, 151
nenko system, 39
Wages and Welfare Department, 59
Weekend work, 97
Within-group/out-of-group distinctions in Japan, 51, 78–79
Womack, James P., 11
Women in work force, 42, 98, 153, 244*n*2
Women's movement in Japan, 48–49
Work Environment Measurement Law of 1975, 67–68
Workers
commitment of, 134
disparate treatment of Japanese, 226
responsibility for safety and health, 28, 70–72, 104, 149
at Jidosha USA, 168, 171–72
See also Labor practices
Workers' compensation, 56, 58, 68–70, 135, 180
benefits, 199–200
premiums, 198–99
Work force
aging of, 74–75
homogeneity or diversity of national, 204
Work hours, 9
Work incapacity resulting from injury vs. illness, 197–98
Working Environment Measurement Law of 1975, 61
Workmens' Accident Compensation Insurance Law, 58
Workmens' Accident Relief Law of 1931, 56–57
Workmen's Accident Relief-Liability Insurance Law of 1931, 56–57
Workmen's Compensation and Insurance Law of 1947, 36
Workmen's Pension Insurance Program of 1941, 34, 57
Work-rule compliance, 77, 139, 189, 229–30
Work shifts at Jidosha USA, 154–55
Workshop, monitoring of, 116
Work teams, 101, 111, 149–50
Workweek, length of, 67, 73, 97–98, 142
World War II, decline in Japanese workers' rights and working conditions during, 34

Zaibatsu (cartels), 35
Zero-accident campaigns, 27, 76, 107, 108, 141, 161, 203

About the Author

Richard E. Wokutch received his bachelor's and doctoral degrees from the University of Pittsburgh and is currently Professor of Management at the R. B. Pamplin College of Business at Virginia Polytechnic Institute and State University. As a Visiting Fulbright Research Professor, he has studied at both Hiroshima Institute of Technology and The Science Center, Berlin.